Logistics and Transportation

Logistics and Transportation

Design and planning

Raja G. Kasilingam

Director of Operations Research/Service Design
CSX Transportation
Jacksonville
Florida
USA

KLUWER ACADEMIC PUBLISHERS
DORDRECHT / BOSTON / LONDON

A C.I.P. Catalogue record for this book is available from the Library of Congress

ISBN 0 412 802902

Published by Kluwer Academic Publishers,
P.O. Box 17, 3300 AA Dordrecht, The Netherlands.

Sold and distributed in North, Central and South America
by Kluwer Academic Publishers,
101 Philip Drive, Norwell, MA 02061, U.S.A.

In all other countries, sold and distributed
by Kluwer Academic Publishers Group,
P.O. Box 322, 3300 AH Dordrecht, The Netherlands.

Printed in Great Britain

Contents

Preface

> If we could first know where we are, and whither we are tending, we could then better judge what to do, and how to do it.
>
> *Abraham Lincoln, 1858, the opening sentence of his 'house divided' speech*

Logistics is nearly a $700 billion industry in the United States of America and is the second largest employer of college graduates. Logistics costs account for nearly 30% of the sales dollar, and logistics activities are essential to satisfying the everchanging customer demand in terms of variety and availability. Shifts in consumer demand patterns, increasing pressure on companies to cut costs, the globalization of several large corporations and the emerging information and communication technologies have created the need for cutting edge, sophisticated logistics practices. The globalization of economic activity has been one of the most important and challenging developments that is having a significant impact on the industries of the developed countries. This has forced major corporations to identify partners in other countries to manufacture components, subassemblies and, in some cases, even the final products. The selection is often based on manufacturing and logistics costs within and between countries. This phenomenon of 'global logistics' has increased the importance of logistics even further.

Logistics encompasses several functions such as vendor selection, transportation, warehousing and facilities planning and location. In addition it is also impacted by production, marketing and product design decisions. Owing to the many interrelated functions and interfaces it is difficult to understand and measure the true logistics costs and to design effective and efficient logistics systems. The requirements or demands placed on logistics functions have been changing over the past several years. These changing requirements may be grouped into the following three categories: competitive pressures, deregulation of transportation industry, information technology. In recent years, logistics has assumed a vital role in satisfying the customer demand at the lowest possible cost. As a result, several new concepts, methodologies and decision support systems have come into existence to meet the logistics challenges of the twenty first century.

The scope of this book is limited to the essential activities within the functional areas of logistics and transportation. The book emphasizes the quantitative treatment of the design and planning issues in logistics. The purpose of the book is to present some of the commonly used decision models and algorithms to address various logistics challenges, the latest trends in logistics and a few real-life examples. The book is very comprehensive covering almost all the elements of a supply chain. It also includes a few topics that are generally not covered by some of the most popular logistics books. These include functional areas such as vendor selection, inventory models with logistics costs, advanced transportation models, logistics metrics and recent trends in logistics. Recommended references for further reading and some of the related training aids are given at the end of each chapter. Two logistics cases are presented at the end of the book.

Chapter 1 provides an introduction to logistics and transportation. Chapter 2 presents the process of analysing logistics systems, and Chapter 3 discusses the concept of integrated logistics network planning. Chapters 4 through 8 describe the various logistics functions and present models and algorithms to address key logistics decisions in these functional areas. Chapter 9 provides an extensive treatment of metrics to measure performance in logistics. The last chapter presents some of the recent trends in logistics such as third-party logistics, benchmarking, virtual warehousing and global logistics. The material covered in this book has gone through classroom testing at senior undergraduate level for a couple of years. Some parts of the material have been used in a couple of graduate courses as supplementary reading material. I have attempted to modify the material to incorporate the feedback from the students.

The book is designed for use in the classrooms by senior level undergraduate students and first-year graduate students from the departments of industrial engineering, civil engineering and business administration. Students specializing in transportation, marketing, materials management and logistics will find this book particularly useful and interesting. Besides engineering and management students, this book will also be very useful as a reference or handbook to practicing engineers, managers and directors from the transportation industry and from the departments of inventory control, material handling, warehouse operations, logistics planning and transportation departments of other industries. The examples and problems used in this text may serve as guidelines to model and solve real-life transportation and logistics challenges.

Raja G. Kasilingam
Jacksonville, FL
October 1997

Acknowledgements

I would like to thank Mark Hammond of Chapman & Hall (UK) for approving my proposal to write this book. I also appreciate his patience and tolerance during the preparation of the manuscript. I would like to thank the Council of Logistics Management for providing me with the electronic copy of the cases presented in this book. A special note of love and gratitude goes to my wife, Dhana and my daughters Aarthi and Preethi, for being very patient and understanding during the entire period of writing this book. Finally, I would like to dedicate this book to my mother and my deceased father.

CHAPTER 1

Introduction to logistics and transportation

1.1 INTRODUCTION

Logistics represents a collection of activities that ensures the availability of the right products in the right quantity to the right customers at the right time. Logistics activities serve as the link between production and consumption and essentially provide a bridge between production and market locations or suppliers separated by distance and time. This requires focus on products or physical goods, people and information about goods and people. Different values are added to a product at various stages of its life cycle. Production and manufacturing adds form value by converting the raw material or components into components or finished parts. Place value is provided through transportation by moving the product where it is needed. Time value is provided through storage and inventory control ensuring the availability of the product when needed. Finally, possession value is added to the product through marketing and sales. Place and time values are added by some of the key logistics functions which are discussed in detail in section 1.2. Example 1.1 demonstrates the value of time and place.

It is to be noted that Example 1.1 does not take into account inventory costs since it assumes known, deterministic demand. A matrix representing the value of logistics based on time and place values is shown in Figure 1.1 (Wilson, 1996). When a product is available at the right time at the right place, its logistics value is at its highest. When it is made available at the wrong place at the wrong time or at the wrong place at the right time, the logistics value is zero. When it is available at the right place at the wrong time, the logistics value is significantly less.

The Council of Logistics Management defines logistics as the process of planning, implementing and storage of raw materials, in-process inventory, finished goods and related information from point of origin to point of consumption for the purpose of conforming to customer requirements. The process of logistics is essential and applicable to all types of industries: manufacturing, retail, health care, transportation and chemicals. Some of the important functions within logistics are

EXAMPLE 1.1

Jarvisons is a local store that specializes in gift, party and seasonal items. The store placed an order for 200 Christmas trees to arrive a week before Christmas in order to meet the expected demand of 200 trees. Owing to a combination of breakdown of the tractor-trailer and rough weather the shipment arrived only on the 26th of December. The regular unit cost transportation and handling is \$5 and the pre-Christmas selling price for the tree is \$30. Owing to the late arrival of the shipment, the store has now to sell the trees at the after-Christmas sale price of \$15/unit. Determine the cost of logistics.

$$\text{profit lost per tree} = \$15$$

$$\text{total profit lost} = \$3000 = \text{cost of logistics}$$

Now, let us suppose that when the trailer broke down, Jarvisons had an option to use another form of transportation to ensure on-time delivery of the shipment. However, the additional cost for this premium transportation was \$10 per tree. What is the value of logistics?

$$\text{lost profit without using premium transportation} = \$15/\text{tree}$$

$$\text{cost of premium transportation} = \$10/\text{tree}$$

$$\begin{aligned}\text{value of premium transportation} &= (\$15 - \$10) \text{ per tree} \\ &= \$5 \text{ per tree}\end{aligned}$$

$$\text{logistics value} = \$5 \times 200 = \$1000$$

vendor selection, inventory control, storage, intra-facility material handling, selection of warehouse and plant locations, layout of facilities and transportation. The overall logistics chain may be divided into three segments: inbound logistics, intra-facility logistics and outbound logistics. Figure 1.2 shows the relationship between the various functions within the logistics supply chain. Some of the logistics functions belong to one of the segments whereas others may span over multiple segments of the supply chain. A supply chain for a particular company may have all or some of the entities (represented as boxes in Figure 1.2) and some or all of the links. A retail store may not have the entity 'plant' and the associated links. For a manufacturer, the entity 'warehouse' may not exist if raw materials and components are shipped directly to the plant. The entity 'distribution center' may consist of a cluster of distribution centers in a hierarchical manner. There may be national, regional and local distribution centers.

Right Place *Wrong Time* Christmas trees in USA but after Christmas	*Right Place* *Right Time* Christmas trees in USA during Nov. 25th to Dec. 24th
Wrong Place *Wrong Time* Christmas trees in Iraq after Christmas	*Wrong Place* *Right Time* Christmas trees in Iraq Nov. 25th to Dec. 24th

Figure 1.1 Matrix representation of time and place values. Source: Wilson, 1996.

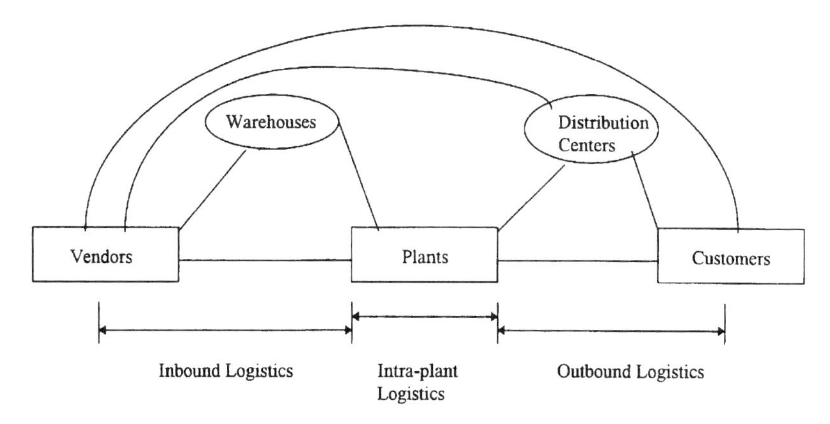

Figure 1.2 Logistics supply chain.

Since the planning and operational decisions of one segment typically has an impact on the other segments, logistics decisions are usually based on a total cost approach. For instance, inventory and transportation decisions are closely related. A lower inventory level may be maintained if a premium or faster transportation mode is used. On the

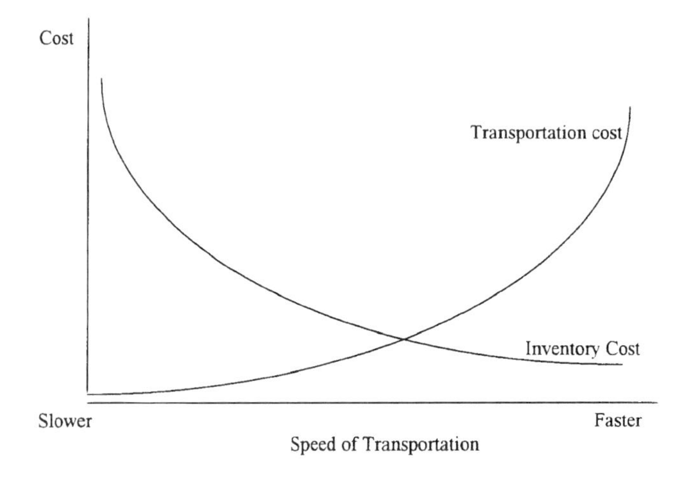

Figure 1.3 Inventory versus transportation costs.

other hand, if a slower mode is used then a higher level of inventory may be needed. Hence a trade-off is needed to determine the optimal inventory level. Figure 1.3 shows the relationship between inventory and transportation costs. Similarly, location and transportation mode and carrier decisions impact one another. Location costs may be lower in terms of acquisition cost, taxes, insurance, etc. but transportation rates may be higher. Another example is the manufacturing of goods in developing countries and shipping them to the point of consumption in developed countries. The trade-off here is between lower raw materials and production costs and higher transportation and storage costs.

Example 1.2 is used to illustrate simply the concept of the total cost approach.

EXAMPLE 1.2

ABC Corporation can purchase a critical component from either of two vendors. The selling price from vendor A is $4/unit, and that from vendor B is $4.1/unit. Based on selling price alone, vendor A is a better option. Now suppose ABC Corporation relocates to a new city and the vendors A and B now charge $0.3/unit and $0.15/unit, respectively, for shipping. Based on total cost, vendor B is more attractive. We may have to consider some other issues besides purchase price and transportation, such as reliability of transportation, quality of supplied parts, ability to supply in small lots etc.

The dependency relationship between the various logistics decisions warrant a total cost approach. In most cases, it may not be possible to consider all the costs and make all the decisions at the same time. In other words, the design of the entire supply chain may be difficult because of the size and complexity of the supply chain. However, some of the very closely related decisions may be made together. Others may be done by using a hierarchical approach. Logistics functions and activities also interface and sometimes overlap with some other business functions such as marketing, finance and production. This aspect is discussed in detail in section 1.3.

The primary objective of logistics is to provide world-class customer service at the minimum possible total cost. Customer service is measured by the availability of the required product of acceptable quality within a specified time. Total cost includes the costs incurred to perform all the logistics functions in order to provide world-class customer service. A recent survey on logistics costs and customer service (Davis and Drumm, 1996) indicates that logistics costs substantially increased from the year 1995 to 1996 to 8% of sales and $80.5/cwt. Only 20% of the companies were able to reduce logistics costs in 1996. The primary reason for the increase may be attributed to more stringent customer service standards requiring higher inventory levels and the use of expensive value-added services. The survey also revealed the following findings:

- logistics costs increased in most industrial sectors;
- most companies were not able to improve customer service levels;
- the number of warehouses seem to decline continually;
- inventory turnover ratio has not improved in general;
- the largest increase was in the warehouse related costs.

The survey demonstrates that there are three major cost drivers that are outside the control of the logistics management of a company – product group, product value and company size. There appears to be a large variation in logistics cost between companies in the same product group and between groups. Logistics costs as a percentage of sales vary inversely with product value. Larger companies have lower logistics costs than smaller companies as a result of economies of scale and the ability to negotiate better rates for third-party services. The survey also shows that the average company fared quite well in customer service – delivering orders accurately, promptly and completely. Customer service as measured by total cycle time, transit time, product availability, complaints, error rates and freight claims were all at their all-time best. These performance levels are being pushed to the next highest level every year by world-class performers.

Table 1.1 Key logistics functions and activities

Logistics function	*Activities/decisions*
Purchasing	Vendor selection, order processing, order follow-up
Inventory control	Order quantity, ordering frequency, inventory valuation, inventory disposal
Facilities location and layout	Number and location of facilities, layout of components within a facility
Transportation	Fleet sizing, routing and scheduling, crew planning, hub or break-bulk terminal location, mode and carrier selection
Intra-facility logistics	Selection of material handling equipment, capacity planning, path design for automated guided vehicles, warehouse design in terms of location and space for items, order picking rules

1.2 SCOPE OF LOGISTICS

Logistics includes all the functions that are essential to provide place and time value to a product. This includes all functions that are necessary to move a product from point of production to point of consumption safely and efficiently. Each of these functions may have certain activities associated with it. For instance, purchasing includes vendor selection, order processing and order follow-up. The key logistics functions and the associated activities are presented in Table 1.1.

1.2.1 Purchasing

Purchasing includes all the activities that have to be performed in order to ensure the availability of materials on time. One of the major challenges for today's purchasing managers is selecting the right vendor(s) for a raw material, component, part or product and determining the amount of order to be placed on each vendor. The quality of parts received from vendors and the timeliness of supply have a significant impact on the ability of a company to meet the demands of its customers completely, accurately and in a timely manner. Selection of vendors is typically based on several important and conflicting criteria such as price, quality, delivery time and service. There are several other factors that influence the selection of vendors. A detailed discussion of the factors and some of the analytical tools available for vendor selection are discussed in Chapter 4. Other important activities of the purchasing function include order processing, following up the orders and rating of vendors based on their past performance.

1.2.2 Inventory control

Inventory planning and control decisions typically follow vendor selection. In some cases they may be made simultaneously with vendor selection. Inventory control decisions focus on the order quantity and the timing between orders. This is done based on lead time, ordering cost, inventory carrying cost, transportation cost, shortage cost, in-transit inventory carrying cost and the level of service in terms of allowable inventory or shortage. The objectives are to minimize the total cost and provide maximum customer service. Since these two are often conflicting in nature, an economic trade-off is needed between inventory levels and customer service levels. Inventory control decisions in a supply chain include the number of stocking locations, product mix at stocking points and the type of inventory strategy – just-in-time, push or pull. Analytical models to address some of the inventory control decisions are presented in Chapter 5.

1.2.3 Facilities planning

Facilities planning addresses two major logistics decisions that are generally made at the initial stages of planning and designing a logistics system: facilities location and facilities layout. Layout and location of facilities play a vital role in minimizing the total cost of logistics. The location of facilities has a huge impact on land and construction costs, local taxes and insurance, labor availability and costs and on the costs of transportation to and from other facilities. The number, size and location of the facilities have a significant impact on inventory-related costs and customer service levels. The layout of a facility has an impact on intra-facility logistics costs such as material handling costs and the costs of material handling equipment. Mathematical and heuristic approaches to solving facilities location and layout problems are presented in Chapter 6.

1.2.4 Intra-facility logistics

This is concerned with the material handling within a large facility such as a plant or a warehouse. Intra-facility logistics is influenced by layout, material handling equipment, stock locations in the warehouse, operating rules for material handling equipment movement and order picking strategies. Typically, a part spends almost 50% of its manufacturing time in moving between machines and storage. There are several important planning and operational problems that need to be addressed to minimize intra-facility logistics costs. The first and foremost is the selection of material handling equipment in order to meet the material handling requirements in terms of weight, volume and frequency of movement and size, value and packaging of the item. The next step is to determine the amount of equipment of each type required and its location if it is fixed-position equipment. If it is flexible-path equipment, then a route structure or guide path must be designed. The operational problems include scheduling of parts and materials to material-handling equipment and vehicles and assigning different types of vehicles between stations. Some of the above mentioned issues are discussed in Chapter 7.

1.2.5 Transportation

Transportation costs typically account for 2.88% of sales (Davis and Drumm, 1996). They are the largest component of logistics cost and represent 40% of that cost. Transportation includes both inbound movement from the sources of raw materials or parts direct to plants or through warehouses and outbound movement of finished products or components from plants to customers directly or through distribution centers. Transportation encompasses a wide spectrum of planning and operational problems. Some of the important planning problems include fleet sizing, vehicle routing, crew planning, network design and hub and terminal location. Crew and vehicle scheduling, dispatching and reservation control are some of the operational problems. Sizing of transportation resources such as trucks, locomotives, cars, aircraft and boats and vessels come under the umbrella of fleet sizing. Vehicle routing focuses on the determination of optimal routings for the various origin–destination traffic, considering route structure, distances and route capacity. Selection of transportation mode and carrier is part of the routing plan but is most often done separately to manage problem size and complexity. Crew planning involves the determination of the staffing requirements to meet the overall fleet operating plan. Network design typically includes the development of routes, schedules and transportation modes and determining the hub locations. Again, it may have to be solved in a hierarchical manner because of the size and complexity of the problem. Procedures and models to address some of the above problems are presented in Chapter 8.

1.3 LOGISTICS INTERFACES

Logistics functions and activities have a significant amount of overlap with the other functions and activities of a company. The functions with a high degree of overlap are finance, marketing, information technology and production. Figure 1.4 shows the relationship between logistics and other business functions. Each function box contains the activities within the function that is impacted by or which impact on

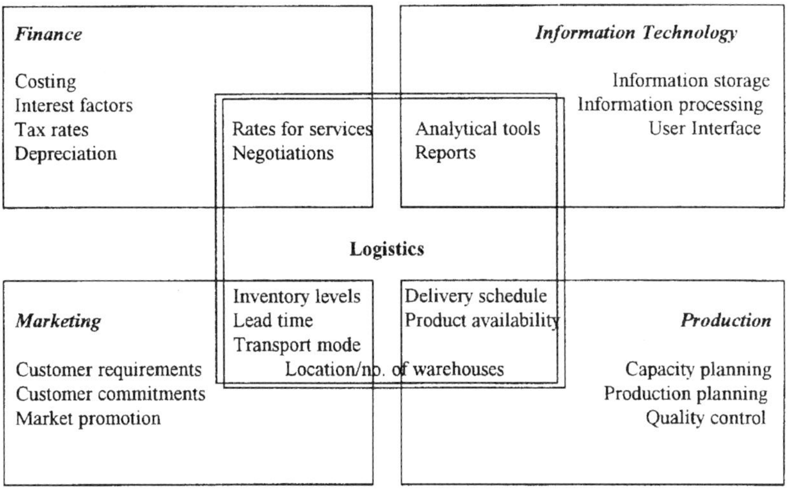

Finance

Costing
Interest factors
Tax rates
Depreciation

Information Technology

Information storage
Information processing
User Interface

Rates for services
Negotiations

Analytical tools
Reports

Logistics

Inventory levels
Lead time
Transport mode

Delivery schedule
Product availability

Location/no. of warehouses

Marketing

Customer requirements
Customer commitments
Market promotion

Production

Capacity planning
Production planning
Quality control

Figure 1.4 Interfaces between logistics and other functions.

logistics. For instance, customer service levels in marketing dictate inventory levels and transportation mode. Or, inventory levels and transportation options impact the level of customer service that can be provided.

1.3.1 Finance

The major link between finance and logistics is through cost data. Cost data provide the basis for making all logistics decisions. Evaluation of alternative logistics systems as well as the development of operating strategies require fixed and variable cost information. The pricing of transportation and storage services requires detailed cost data on fuel, maintenance, handling, labor, material and overheads. Equipment purchasing and replacement decisions cannot be done without information on depreciation methods and tax structure. Negotiation of rates or prices with suppliers and third-party providers cannot be done without having accurate cost data.

1.3.2 Marketing

Marketing provides the necessary information that serves as the starting point for designing the logistics system. It includes customer requirements in terms of product variety and response time, customer commitments in terms of time, quality and quantity and other market promotions. Almost all of the logistics decisions are made in order to meet customer commitments and requirements – number and location of stocking points, variety and quantity of stock, transportation mode options and type of packaging. The logistics chain must be responsive to the promotions initiated by the marketing department in terms of time, place, quantity and product variety.

1.3.3 Information technology

This is a function that has been gaining importance across all the functions in a company. This increase in importance is primarily a result of the availability of software and hardware at affordable prices as well as the advancement in hardware and software in terms of handling a large volume of data and ease of usage. Most logistics functions require the storage, processing and retrieval of enormous amounts of data, real-time communication capabilities, a simple and easy-to-use user interface and the ability to support sophisticated analytical tools and report generators. Some of the recent developments in information technology can meet all of the requirements of the logistics function.

1.3.4 Production

Whereas marketing provides the customer requirements, it is production that ensures that the customer requirements are met in terms of quality, features, quantity and timeliness. The product design group translates the customers' wish list into reality by selecting suitable materials, colors and determining the appropriate dimensions. Process planning translates product design into a series of manufacturing steps by identifying the type of machines, tools and fixtures required at each step. Capacity planning determines the required number and type of machines to manufacture the desired part. Production planning schedules the manufacturing of the part by balancing the use of available resources (machines and labor) and inventory on hand to meet the delivery time and quantity required.

1.4 LOGISTICS PROFESSION

This section focuses on two aspects. The first one is concerned with the career choices available in logistics and transportation. It covers the type of positions in logistics and the associated responsibilities, the background required for these positions in terms of education, training and experience and the career path in logistics. The second aspect focuses on the organizations that are primarily devoted to serving logistics professionals. A list of addresses is given at the end of this chapter.

1.4.1 Careers in logistics

The publication of the Council of Logistics Management entitled *Careers in Logistics* (1996) lists seven different logistics job profiles. The profiles are inventory control manager, warehousing/operations manager, administrative manager, administrative analyst, transportation manager, customer service manager and consultant. The publication also presents results of interviews with people working in the above job profiles. The results include how they began their logistics career, their primary and secondary job responsibilities, the types of skills and knowledge needed for their job and the amount of time they typically spend on analytical work, meetings, presentations and supervision. There are also other logistics job profiles that are very specific: logistics/transportation network service manager, car manager, fleet manager, transportation network planning manager, network operations manager, routing/scheduling manager, facilities planner, materials manager, purchasing manager, inbound logistics controller, warehouse manager, logistics system analyst and so on. Some of the titles are unique to the type of industry, such as transportation, retail and electronics. Within the same industry, different sectors (trucking, rail, air and container sectors in transportation) or different companies within the same sector may have jobs with similar responsibilities with different titles or with similar titles with different responsibilities.

Regardless of the type of industry or business, every company will have some type of logistics functions, be it in the transportation or health care industry. Every company will have some type of commodity (people, information or products) that needs to be transported and/or stored. Hence the potential employers of logistics professionals include almost all types of industries. The responsibilities, nature of job and the functional areas of coverage may vary based on the position within a company and the size of a company. Generally, lower level positions involve more analytical work and focus on one of the logistics functions. Higher level positions involve the directing, planning and control of more than one logistics function. However, a logistics position in a smaller company may provide the opportunity to have a combination of analytical and supervisory responsibilities as well as be exposed to several functions. The typical career progression is from analyst to manager to director, and then to vice president and senior vice president of logistics and/or distribution or chief officer of transportation, logistics or distribution.

The requirements to begin a career in logistics vary depending upon the functional area, type of industry and company size. The most basic requirement is a college degree in engineering, business administration, finance or in similar fields. Most companies also prefer candidates with some experience either through summer internship programs or through co-op programs. Several companies provide opportunities for internship or co-op to students during their undergraduate education. It is also becoming increasingly common for companies to seek candidates with advanced degrees. This is especially true in the areas of logistics planning and design which involves quantitative analysis. Candidates with advanced degrees generally start at higher level positions and receive higher starting salaries. This may not be true for people who have been in the industry for a while.

There are also certain skills that are essential to perform most of the logistics functions. Good computer skills are a must – an ability to use spreadsheets, databases, word processors, and the possession of some programming capability. Oral and written communication skills and people skills are essential in logistics as in any other field today because of the globalization of industry and competition. During undergraduate or graduate education it is highly recommended to obtain some background in cost and economics, statistics and quantitative models, operations management, transportation and warehousing, communications, marketing and computer science. Most of the logistics and transportation degree programs are offered through business and management schools. There are also very good programs that are offered through engineering schools. The Georgia Institute of Technology and the University of Arkansas are a few to mention in the USA that belong to this category.

A summary of a recent survey of career patterns in logistics (La Londe and Masters, 1996) is given below.

- The career patterns in logistics continue to remain stable. Employment opportunities for new college graduates and experienced executives are improving as more companies are placing a higher emphasis on logistics functions. Salary levels of beginners as well as of experienced logistics professionals are increasing. Logistics professionals are working long, hard hours, and are increasingly using personal computers.
- Functional responsibilities of logistics professionals also remain stable. Most logistics professionals are employed in the areas of traffic, warehousing and inventory management. Only very few are active in the areas of forecasting, product planning, purchasing and packaging.
- Most logistics professionals see information technology as a high priority in their continuing education. They also see information technology as the primary tool they will use to solve logistics problems and capture other logistics opportunities in the future.

1.4.2 Professional logistics organizations

Logistics costs account for a significant part of the total cost of a product. Owing to the increasing importance of logistics functions, a number of professional organizations are dedicated to logistics, transportation and inventory management. The primary objective of these organizations is to advance the professional knowledge of their members through conferences, seminars, continuing education programs, publications, professional certification opportunities and local chapter meetings. The organizations also constantly strive to bridge the gap between research and practice by bringing together academics and industry experts.

Council of Logistics Management

The Council of Logistics Management (CLM) is a premier organization and probably the most widely known organization for logistics and transportation. This was formerly known as the National Council of Physical Distribution Management. In 1997, the CLM had nearly 13 000 members from all over the world. The CLM organizes an annual conference which is attended by about 5000 professionals. The conference has a separate day of sessions for the educators as well as tours to the facilities of companies with cutting edge logistics practices. The *Journal of Business Logistics* published by the CLM is considered as one of the leading international journals in logistics. The CLM has about 53 roundtables worldwide which are involved in different types of activities such as awareness programs, fund raising and knowledge exchange.

American Society of Transportation and Logistics

The American Society of Transportation and Logistics (AST&L) is a professional organization with a primary focus on transportation and physical distribution. The AST&L organizes a transportation education symposium annually to bring practitioners up to date with the latest developments in transportation and logistics. It also offers professional certification in transportation and logistics. Becoming a certified member requires passing four comprehensive tests covering major areas of traffic, transportation and logistics and writing an original research paper in any of the transportation and logistics areas. *The Transportation Journal* published by AST&L deals with some of the non-quantitative sides of transportation such as legal and policy issues and surveys. The AST&L also publishes a newsletter which periodically includes a list of positions available in transportation and logistics.

The International Society of Logistics

The International Society of Logistics is an international organization devoted to scientific, educational and literary endeavors to enhance the art and science of logistics technology, education and management. It was formerly known as the Society of Logistics Engineers. It has about 7500 individual members and 25 corporate members.

American Production and Inventory Control Society – The Educational Society for Resource Management

This was founded in 1957 as the American Production and Inventory Control Society (APICS). Since then it has evolved to meet the changing needs of business by providing broad-based individual and organizational education focused on integrating resources for improved productivity. APICS offers a full range of cost-effective, results-oriented education options for the manufacturing and service sectors, including conferences and seminars, books and publications and professional certification programs.

Other organizations

There are several other logistics, transportation and inventory management organizations that serve the needs of logistics professionals. The Eno Transportation Foundation Inc. based in Virginia, USA, The Institute of Logistics based in Northants, UK, the Transportation Research Board based in Washington, DC, USA and the Transportation Research Forum based in Virginia, USA are some of the prominent organizations in transportation and logistics. The International Society for Inventory Research based in Lund, Sweden and the National Association of Purchasing Management based in Arizona, USA focus on supplier selection, purchasing and inventory control. The Warehousing and Education Research Council (WERC) based in Illinois, USA concentrates on storage and warehousing related activities. The objectives of WERC are to provide education and to conduct research concerning the warehousing process and to refine the art and science of managing warehouses.

1.5 ORGANIZATION OF THE BOOK

In this book, I have made an attempt to discuss and analytically address some of the key logistics functions. The objectives of the book are to provide a reasonable insight into the various functions, to identify some of the important decisions in each of the functions and to present some quantitative methodologies to help make the decisions. Accordingly, the first two chapters of the book provide an insight into the overall logistics design process and the subsequent chapters are organized by functional areas. Chapter 2 presents data collection and analysis and modeling approaches to performing logistics system analysis and design. The changes in certain factors that lead to the design or redesign of logistics systems are first discussed. The four important phases of the process of logistics systems – problem definition, data analysis, problem analysis and system implementation – are then presented. Problem definition focuses on the level of detail, system objectives and the type of logistics product. Data analysis addresses data collection, sample size, outlier screening, data aggregation and estimation of missing or unavailable data. Guidelines for logistics system design and various types of models for developing and selecting the best design are presented as part of problem analysis. System implementation covers validation, verification, sensitivity analysis, user training and user testing.

Chapter 3 presents a network-based approach to logistics system or supply chain planning. The importance of network-based planning is presented first. The various components of the network and the relationship between costs and operating levels of the network components are discussed. Then a generic network formulation for designing the supply chain is presented.

Chapter 4 discusses the importance of vendor selection and several approaches to vendor selection under a variety of situations. The first part of Chapter 4 presents some of the important criteria for vendor selection and a classification of vendor selection decisions. The second part covers vendor selection methods, including empirical methods which include factor analysis, vendor profile analysis and the analytic hierarchic process and optimization methods, which encompass the application of facility location models and a model based on quality costs.

Chapter 5 provides an insight into integrating inventory and transportation issues while addressing inventory control decisions. The important functions of inventory and the different costs of inventory are presented first, followed by a classification of inventory control problems. An extension of the classical economic order quantity (EOQ) to include transportation and in-transit inventory costs is presented. Then a linear programming formulation and transportation model formulation of the multiperiod inventory planning problem are presented. Three different models to represent just-in-time ordering situations are presented – the first one determines the frequency of delivery for blanket ordering, the second considers order quantity and frequency of delivery and the third model combines vendor selection and just-in-time inventory control.

Chapter 6 addresses facility location and layout decisions. It begins with a classification of facility location problems. Single and multifacility location models under Euclidean and rectilinear distance functions are covered. A methodology based on dynamic programming is presented to address facility location over a planning horizon consisting of several periods. A section is devoted to address popular facility location models such as the simple plant location model, the quadratic programming model and set covering and partition models. Finally, three different approaches to facilities planning are addressed – the systematic layout planning process, an improvement-based heuristic and a construction-based method.

Chapter 7 discusses the importance of intra-facility logistics or material handling. The different types of material handling equipment are covered first. Then the selection of material handling equipment based on functional and financial requirements is presented. Some of the planning and operational problems of automated guided vehicles are also addressed.

Chapter 8 covers some of the major transportation decisions and several analytical models used to address the decisions. An introduction to various types of transportation systems is presented first, followed by a discussion on transportation costs, economics and rates. A broad range of simple operations research models such as transportation, the trans-shipment model and the traveling salesman model are presented. Some real-world transportation problems from airlines, railroad, trucking and surface transportation are also presented. Operations research modeling of the following problems are presented: fleet sizing, hub location, locomotive planning, railroad blocking, crew scheduling and cargo revenue management.

Chapter 9 presents the importance and an overview of some of the key logistics performance metrics. A high-level classification of logistics metrics is provided based on the following: internal versus external, financial versus non-financial and system versus functional or unit level. Several examples of the different types of metrics are also given. Finally, some of the important issues in logistics metrics such as criteria for developing metrics, units of measurement, reporting process and understanding the drivers of metrics are discussed.

Chapter 10 outlines some of the recent trends in logistics such as reverse logistics, global logistics, benchmarking and virtual warehousing. The importance and necessity for following these trends in an appropriate manner are discussed. Some of the tools and techniques available for executing these trends are also discussed.

1.6 SUMMARY

Several functions of logistics have been around with several companies for a long time. Increasing importance has been placed on logistics in the 1990s and it has become a new and emerging area of focus in several large corporations. In the twenty first century it will demand more focus and attention of the senior executives because of the potential opportunities for cost reduction and improvement to customer service. The increase in the number of organizations that serve the logistics and transportation professionals and the growth in membership of these organizations worldwide indicate the importance of logistics. This is supported by the fact that the number of research journals, magazines, seminars, short courses and conferences in logistics has been on the increase over the past several years. This chapter has highlighted the importance of logistics and the various functional areas of logistics. It has also presented some of the other business functions that interface with logistics. The careers available in logistics and some of the professional organizations in logistics were also discussed. A chapter-by-chapter overview of the contents of the book was also presented toward the end.

PROBLEMS

1.1 Discuss briefly the major functional areas within logistics.

1.2 Explain the types of value provided by logistics by using (a) an example from manufacturing, (b) an example from retail industry.

1.3 What are the different career choices available to logistics graduates? What are the associated responsibilities?

1.4 Why is total cost approach very critical in making logistics decisions? What are the barriers to such an approach?

1.5 Research your library and list the various journals and magazines in the different areas of logistics.

1.6 Identify some of the local chapters at your university and in the city that serve logistics students and professionals.

1.7 A store in New York placed an order for a shipment of 1000 roses from Mexico for Valentine's day. Owing to some problems en route, the shipment arrived a day after Valentine's day. The total cost of transportation was $800. The store ended up selling the roses for $1 less than the Valentine's day price. What is the cost of logistics?

1.8 ABC Company located in Fayetteville, AR, purchases its raw material from Memphis, TN. The company has options to use carrier A and/or carrier B. Carrier A charges $2.00/cwt and carrier B charges $2.20/cwt; however, B has a quantity discount option which will reduce the rate to $1.80/cwt if a transportation volume of 10 000 lb a month is guaranteed. The following is the production plan of ABC company for the next three months:

Month	1	2	3
Production (finished units)	1200	700	1100

Each unit of finished product requires 10 lb of raw material. Assuming unlimited storage space at the plant site and a company policy to order just to cover monthly requirements, determine a single carrier to use for the next three months.

1.9 SmartShirts, Ltd. has two production facilities, one in Chicago and another in Atlanta. It has the option to buy cloth or fabric from a supplier in Los Angeles or New York. The fabric cost per shirt is $4 if purchased from Los Angeles and $4.3 if purchased from New York. The transportation cost per shirt material from Los Angeles to Chicago is $1.2 and from Los Angeles to Atlanta is $1. The transportation costs from New York to Chicago is $0.8 and from New York to Atlanta is $0.7. Production costs are $3 per shirt in Chicago and $3.2 per shirt in Atlanta. Assuming unlimited production and transportation capacities, what will be the best alternative to produce shirts?

REFERENCES

Council of Logistics Management (1991) *Careers in Logistics*, 2803 Butterfield Road, Suite 380, Oak Brook, IL 60521.

Davis, H.W. and Drumm, W.H. (1996) Logistics costs and customer service levels. *Council of Logistics Management Annual Conference Proceedings*, Council of Logistics Management, 2803 Butterfield Road, Suite 380, Oak Brook, IL 60521, pp. 149–59.

La Londe, B.J. and Masters, J.M. (1996) The Ohio State University Survey of Career Patterns in Logistics, *Council of Logistics Management Annual Conference Proceedings*, Council of Logistics Management, 2803 Butterfield Road, Suite 380, Oak Brook, IL 60521, pp. 115–138.

Wilson, J. (1996) Quantifying value creation across the logistics channel. *Council of Logistics Management Annual Conference Proceedings*, Council of Logistics Management, 2803 Butterfield Road, Suite 380, Oak Brook, IL 60521, pp. 7–113.

FURTHER READING

Council of Logistics Management (1991) *Logistics in Service Industries*, 2803 Butterfield Road, Suite 380, Oak Brook, IL 60521.

Council of Logistics Management (1994) *Bibliography of Logistics Training Aids*, 2803 Butterfield Road, Suite 380, Oak Brook, IL 60521.

Wood, D.F. and Johnson, J.C (1993) *Contemporary Transportation*, Macmillan, New York.

TRAINING AIDS: VIDEOTAPES

Council of Logistics Management, 2803 Butterfield Road, Suite 380, Oak Brook, IL 60521, tel. (708) 574-0985: *Logistics: Careers with a Challenge*.

Penn State Audio-Visual Services:
Business Logistics Management Series: Customer Service;
Business Logistics Management Series: Introduction to Logistics;
Business Logistics Management Series: Internal Logistics Environment;
Business Logistics Management Series: Logistical Relationships in the Firm.

USEFUL ADDRESSES

United States

American Production and Inventory Control Society (APICS) – The Educational Society for Resource Management, 500 West Annandale Road, Falls Church, VA 22046.

American Society of Transportation and Logistics (AST&L), 216 East Church Street, Lock Haven, PA 17745.

Council of Logistics Management (CLM), 2803 Butterfield Road, Suite 380, Oak Brook, IL 60521.

Europe

Institute of Logistics, Douglas House, Queens Square, Corby, Northants, UK.

CHAPTER 2

Logistics systems analysis

2.1 INTRODUCTION

As discussed in Chapter 1, logistics is the process of moving, storing and retrieving material, people and information efficiently and economically. Logistics systems perform all or some of the essential logistics-related functions to achieve the desired objective(s) of efficiency and/or economy. An example of a logistics system may be a distribution system of a retail company. The essential logistics functions in this case include location of warehouses, selection of transportation options, inventory decisions for warehouses and stores, location of consolidation or redistribution points, selection of vendors etc. Some of these functions may be performed by in-house people; others may be carried out by third parties. Regardless, the system has to be designed to carry out these functions in order to meet certain objectives.

The problem of logistics system design arises under two situations. The first case is when a new logistics system has to be designed. The second one is when the existing system has to be redesigned to accommodate certain changes. These changes may occur in any of the following: customer service, demand, product characteristics, costs and pricing policy (Ballou, 1992).

Customer service requirements may change as a result of competition or changes to customer service policy. For instance, competition may force a trucking company to improve its dock-to-dock service reliability (delivering to the customer within x hours of the promised time). Alternatively, a company may decide to set higher targets for service reliability in anticipation of capturing more market share and/or demanding a higher price. Other customer service requirements may be a shorter lead time, availability of products both in terms of time and variety and better warranty service. Change in customer service requirements may require changes in several logistics functions: additional warehouses, change in transportation carrier and/or mode or higher inventory levels. These changes lead to the redesign of an existing system.

Change in demand may be due to a shift in population as well as a shift in consumer attitudes. This causes changes to demand by location, time and product type. This may require the opening of new stores or warehouses in some areas, the closing down of facilities in other areas and changes to stock levels of different items.

Product characteristics change as a result of the functional or aesthetic redesign of the products to meet new markets and changing customer needs. Technological changes result in smaller and lighter products arising from innovations in computer hardware and the invention of new materials. Change in product characteristics such as weight, volume and shape affects packaging, handling, storage and transportation options.

Changes to labor agreements, advances in information systems, technological changes in material handling, storage and transportation, new transportation rates and new customs regulations cause logistics costs to increase or decrease. Since some or all costs have changed, a system that seeks a new trade-off between these costs is needed.

A company may change its pricing policy to reflect the changes in logistics costs. Also, sometimes price may be reduced to increase market share or increased to boost revenues. Change in price may also be effected by changing delivery or transportation terms in the sense of who pays for what expenses. Change in price may have an impact on product demand as well.

There are several key steps involved in designing, developing and implementing an efficient and economical logistics system. The process of logistics systems analysis may be grouped into four phases. The first phase is concerned with defining the problem. This includes defining focus area(s) and system objectives and understanding the basic logistics product. The second phase is related to data analysis. This includes identifying data sources, analysing the adequacy and accuracy of available data, developing methodologies for collecting additional data, developing estimates for unavailable data etc. The third phase consists of analysing the problem in hand using the available or estimated data. Several tools may be used for logistics system analysis. The most commonly used tools are simulation, operations research models and heuristic procedures. These tools enable logistics managers to develop alternative designs and select a final design for the system. The final phase includes user testing and actual implementation of the logistics system. This process is shown in Figure 2.1. Examples and additional discussions of the four phases are given in the following sections.

2.2 PHASE 1: PROBLEM DEFINITION

Definition of the problem is vital to the subsequent design of the system. One has to understand the difference between a symptom and

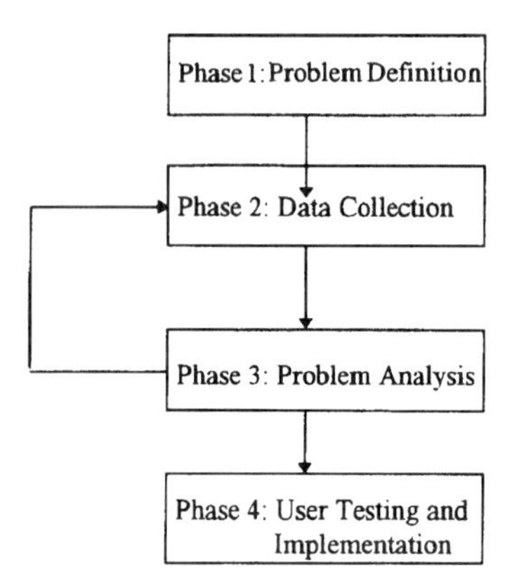

Figure 2.1 Logistics systems analysis process.

a problem. In many cases, we tend to define a symptom as a problem. A good example is an out-of-stock situation of a particular item. On the surface this may seem to be a result of poor inventory decisions. However, the real problem may be unreliable transportation. A problem must be defined at the appropriate level. The definition must include the expectations from the new system, business and operating rules and constraints and the criteria to be used for selecting the best system from among the various alternatives. Above all, the logistics product that flows through the system must be defined. Some of the critical issues in problem definition are discussed below.

2.2.1 Level of detail

When developing a new system or redesigning an existing system a fundamental question is whether to consider the complete logistics system or only a few critical components of the system. The answer to this question depends upon: the availability of resources, the size of the logistics network, the relative magnitude of components and costs and the degree of system integrity. The ideal approach is to consider the entire network. However, in several cases the changes to the system inputs may have only a weak link with certain components; the time and resources available may be limited and the size of the logistics network may be too large to solve as a single system. A more practical approach is to consider the problem in a hierarchical manner and design the individual components separately and then integrate them together. A network-level design approach is discussed and presented in Chapter 3. Approaches to solve the individual components are presented in Chapters 4 through 8.

Level of detail is also related to the type of logistics plan. A strategic plan will require a long-term, macrolevel problem definition. An operational logistics plan will require a short-term, detailed definition of the system. Determining the type of aircraft to serve a market is an example of a strategic problem whereas the assignment of a specific aircraft of that type to serve that market on a given day is an operational problem. Level of detail also depends upon the physical size of the problem. For instance, for a company that operates nationwide the definition of demand center may be at a very aggregate level when addressing the problem of locating national distribution centers. It will be at a detailed level for a company that operates within a region or within the metropolitan area of a city.

2.2.2 System objectives

The business objectives of a company must be clearly established to facilitate the efficient design and development of a logistics system. Business objectives impact the type of logistics strategies which in turn determine the design of the logistics network and/or its components. Three commonly adopted business objectives are: minimize capital cost, minimize operating cost and maximize customer service. The capital cost-reduction objective aims to minimize the total investment in the logistics system. The corresponding logistics strategies may be to avoid warehousing and to ship direct to customers or to select public warehouses over private warehouses. For instance, the logistics network to support this strategy will focus on efficient transportation alternatives and on locating plants closer to markets.

The operating cost-reduction objective minimizes the variable cost of movement and storage. The corresponding logistics strategies may be to select better storage and transportation options. For instance, the use of truckload carriers over less-than-truckload carriers and use of private warehouses. The service-maximization objective maximizes product availability and minimizes order response time. The logistics network to support this will have more warehouses, increased product mix at retail warehouses and premium transportation services.

2.2.3 Logistics product

It is essential to understand the logistics product since it is the commodity that flows through the logistics pipeline. The type of logistics system depends upon the various attributes of the product such as the market of the product, the importance of the product, product characteristics and product packaging.

Market of the product

This distinction is based on the ultimate market of a product. Products that are purchased by individuals or companies to produce other goods or services are known as industrial products. Examples include machine tools and subassemblies of other major end products. On-site service or fast and responsive after-sales service is often critical for these products. Products that are directed for the use of individuals who are ultimate consumers are known as consumer products. Consumer products can be further classified into convenience, shopping and specialty products. Convenience products are purchased frequently for immediate use with little comparative shopping. Examples include food items sold in vending machines, items displayed near checkout counters and items generally available in stores such as Seven Eleven. These products require a wider distribution through many outlets – product availability and convenience are important. Shopping products, on the other hand, are purchased after comparing price, quality and performance. Consumers read product surveys, seek opinions from friends and even test the look and feel of the product before buying. Examples include stereos, automobiles, appliances and furniture. For these products the number of stocking points and the stock held is usually very low. Only floor samples of a few makes and models are held in inventory. For specialty products, consumers are willing to spend a considerable amount of time to decide and own the product. Brand preference is very common. Examples include luxury automobiles, custom home furnishings, etc. In most cases, only catalogue descriptions are available for these products and special orders have to be placed for product availability.

Importance of the product

This indicates the value of the product to the company. Value may be defined in terms of revenue, profitability or contribution of the product. For most companies, a very few products account for a large portion of revenue or profitability. These products are classified as A items. Typically, A items consist of 20% of the products and account for 70%–80% of the revenue. B items consist of 30% of the products and represent 10%–20% of the revenue. Finally, C items consist of 50% of the products and make up only 10% of the revenue. ABC classification is very commonly used in inventory control, the stocking of items in a warehouse and distribution planning. A items require wider distribution with high stock availability, are preferably located near material handling equipment in a warehouse and stock control must be frequent and tight. C items fall at the other end of the spectrum: they have limited distribution channels with low stock levels, they are located far away from the material handling equipment and there is infrequent and relaxed inventory control. B items fall in the middle of the spectrum.

Product characteristics

Important product characteristics include density, the value/weight ratio, substitutability and the risk aspects of the product. Density plays a vital role in determining storage and transportation capacity. Volume capacity is used up before weight capacity when storing or transporting low-density products. For heavy products, weight capacity is used up before volume capacity. In general, transportation rates are weight-based and storage rates are volume-based and value-based. Hence a minimum charge is often stipulated by carriers for very light products in transportation and by warehouses for very heavy products in storage to avoid unusually low rates. The value/weight ratio impacts transportation and storage rates. In general, as the value/weight ratio increases storage rates increase and transportation rates decrease. Density and the value/weight ratio impact logistics system design through their impact on transportation and storage costs.

Substitutability refers to the willingness of the consumers to buy another make or model if the desired model is not available. For a highly substitutable product availability must be ensured at all times by maintaining high inventory levels and by using reliable transportation options for on-time replenishments, otherwise sales will be lost. Risk aspects of the product include perishability, hazard potential and vulnerability to theft. Perishable products need stringent inventory controls and refrigerated transportation and storage conditions. Contaminants and explosive products need temperature control and pose restrictions on stacking height and storage and handling with other items. Small items that can be easily stolen need to have frequent stock control and high security.

Product packaging

Products are packaged for protection and sales promotion. Packaging also facilitates storage, handling and transportation. The type of packaging impacts storage, handling and transportation alternatives and influences costs of storage, handling, transportation, package and damage.

2.3 PHASE 2: DATA ANALYSIS

The second phase is related to data collection. This includes identifying data requirements, data sources, analysing the adequacy and accuracy of available data, data aggregation, developing methodologies for collecting additional data, developing estimates for unavailable data etc. Data requirements depend upon the nature of the project, and data sources depend upon the data needs. For instance, a problem on vendor selection will require data on the location of vendors, vendor capacity, selling price, quality, shipment quantity, lead time, after-sales service, mode and cost of transportation etc. The corresponding data sources may be vendors, transportation carriers and past vendor performance. A problem on warehouse location will require data on site cost, construction cost, relocation cost, storage requirements, distance to material sources and sinks, transportation alternatives and costs, service level desired etc. Data sources may include a market-demand database, rate tables, Chamber of Commerce etc.

Once data sources are identified it is very important to analyse the adequacy and accuracy of the data. Adequacy has two dimensions: completeness and sample size. A data stream is complete when it has no missing data for certain days or months. Data are not complete when some of the data attributes are missing. The term 'attribute' refers to the various pieces of data. For example, daily demand data for freight must include date, weight and volume since freight densities vary by commodity; weight or volume alone is not sufficient. Sample size refers to the number of data points required to have a desired level of accuracy in the analysis. Data accuracy refers to the reliability of data. The accuracy of data can be verified by scanning for extremely large or small values that are outside the acceptable range. Comparison of data from two or more sources for the same data element may also help to validate data accuracy. The values of two or more dependent data elements that do not correlate also indicate inaccurate data. This is best explained by Example 2.1.

EXAMPLE 2.1

The freight demand data for market A for three days is as follows.

Date	*Commodity*	*Weight (lb)*	*Volume (ft^3)*
1 January 1997	1	200	10
2 January 1997	2	400	40
3 January 1997	1	300	10

Clearly, there is a problem with the data for 1st or 3rd January. On both dates freight was made up of commodity 1; but weights and volumes do not correlate.

Data aggregation is helpful in reducing the problem to a manageable size. In logistics systems analysis we may have to aggregate several locations into one or group similar products into one family. Grouping is normally done based on certain characteristics. Grouping of products may be done based on their similarity in product characteristics such as weight, volume, perishability and risk to exposure. Once grouped, products within a group may be handled, stored or transported in a similar manner. Development of estimates for unavailable or missing data may be done by using expert opinion, simple interpolation or simple-to-sophisticated forecasting methods. In certain cases, functions may have to be developed to represent the relationship between data elements, to capture variation of data over time and to understand the uncertainty associated with data. The following sections provide some tools and techniques for data collection and analysis.

2.3.1 Data collection

When data are not available, methods and systems have to be put in place to begin collecting data. This may range from manual collection and entry to automatic scanning and loading into a database. The type of data collection method varies depending upon the amount of data, level of information technology and the limit on capital expenditure allowed for data collection. The encoding of data facilitates data collection and further analysis. Example 2.2 illustrates data encoding.

EXAMPLE 2.2

ABC Airlines is in the business of moving freight within Europe. The company serves several cities, carries different types of commodities and provides varying service levels to the customers. A suitable coding scheme for the company may be as follows.

Service level		*Commodity*	
type	*code*	*type*	*code*
Same day	SMD	Perishable	SCR
Next day	NXD	General	GEN
Overnight	OVN	Express	EXP

City codes will be the same as standard three-character airport codes (for example, FRA for Frankfurt and LHR for London Heathrow).

2.3.2 Sample size

The validity of results from logistics analysis depends upon the sample size or the number of data points used in the analysis. The size of the sample increases with the desired level of accuracy, confidence level and the variability in the data. Level of accuracy defines the deviation of the estimate from the actual value. When the variability of data is not known *a priori* it may be estimated from a reasonably large sample (greater than 30). When our objective is to use the data to compute the mean of a certain variable, the following formula can be used for estimating the sample size n:

$$n = \left(\frac{sZ_{\alpha/2}}{e}\right)^2 \tag{2.1}$$

where,

s is the population variance or its estimate;
e is the desired accuracy;
$1 - \alpha$ is the confidence level;
$Z_{\alpha/2}$ is the standard normal variate.

The desired accuracy can also be expressed as a function of the relative accuracy and the estimate of the mean. Let us say that the mean computed from an initial sample ($n > 30$) is 25 and the relative accuracy desired is 10% within mean, then the desired accuracy is 2.5 (25×0.1). In this case, sample size can be estimated as follows:

$$n = \left(\frac{sZ_{\alpha/2}}{AX}\right)^2 \tag{2.2}$$

where,

A is the relative accuracy;
X is the sample mean.

An illustration of the use of equation (2.2) is given in Example 2.3.

EXAMPLE 2.3

LMT Transportation measures its daily service performance based on the percentage of shipments that are delivered to the customers within two hours of the promised time. Determine the required sample size for computing the average service reliability performance given the following data: $s = 0.3$, $e = 0.05$, and $1 - \alpha = 0.95$. Hence $\alpha/2 = 0.025$ and, from standard normal tables, $Z_{\alpha/2} = 1.96$. Thus, from equation (2.1)

$$n = \left(\frac{0.3 \times 1.96}{0.05}\right)^2 \approx 138$$

Sometimes, we may want to verify if the available number of data points is sufficient for a given confidence level and accuracy. In this situation, compute the acceptable variance, s_{ac}, as follows:

$$s_{ac} = \frac{en^{1/2}}{Z_{\alpha/2}} \tag{2.3}$$

Compute the actual variance from the data. If actual variance is less than or equal to the acceptable variance then the sample size is sufficient.

2.3.3 Outlier screening

Outliers are data points that are not representative of normal data. Examples include a large and very light one-time shipment, a very high rate for a special shipment, railroad performance statistics during the Christmas period etc. When one computes the average shipment density a single large and light shipment will skew the density downward. Similarly, a very high rate for one special shipment will skew the average rate upward. Outliers in data can be screened by using two simple methods: the percentile method and value-based method. In the percentile method, data above and below certain percentile values are eliminated. For instance, all data points with density values less than 5th percentile and more than 95th percentile are considered as outliers and are eliminated. The weakness of this method is that 10% of the data will be eliminated even if they are not outliers. In the value-based method data points with values below a specified value and above a specified value are screened out. An example may be to exclude all density values less than 0.1 lbs/ft^3 and more than 30 lbs/ft^3. The cutoff values need to be specified or known for this method.

2.3.4 Data aggregation

Clustering methods are commonly used for grouping similar products or locations to reduce problem size. The key is to define a similarity measure based on certain attributes. For example, similarities in products may be expressed in terms of size, shape, color, weight and density.

Similarity measures

Definition of similarity measures based on distance metrics is explained below, using products as an example. Each product j can be assigned a set of n attribute values $X_{j1}, X_{j2}, X_{j3}, \ldots, X_{jn-1}, X_{jn}$. The similarity between two products j and k may be expressed by means of the Minkowski distance metric as follows:

$$d_{jk} = \left[\sum_{p=1}^{n} |X_{jp} - X_{kp}|^r \right]^{1/r} \quad (2.4)$$

where r is a positive integer. For $r = 1$, equation (2.4) is known as the absolute metric and for $r = 2$ it is known as the Euclidean metric. There are other definitions of distance metric such as the weighted Minkowski metric and the Hamming distance metric (Singh and Rajamani, 1996). For an illustration, see Example 2.4.

Clustering methods

Clustering is a generic name for a variety of statistical and mathematical models to group similar objects into a set. The most commonly used clustering methods include single linkage clustering, average linkage clustering, complete linkage clustering and linear cell clustering. The basic steps in all the methods are almost the same (Singh and Rajamani, 1996). The complete linkage clustering algorithm is presented in this section.

Step 1: choose the desired number of final groups. Compute the similarity matrix between the various products using any of the distance metrics. Initially assume that each product is in a separate group.

Step 2: find the maximum value in the distance matrix (meaning maximum similarity). Merge the two groups (say j and k) that correspond to this value into a new group, t. Now recompute the similarities between the new group and other groups [remember a group may have only one product or several products] as follows: the similarity between group t and group s is equal to the minimum of {similarity (l, m)} where $l \in t$ and $m \in s$ (the maximum of {similarity (l, m)} in the case of the single linkage clustering method). Remove groups j and k from the matrix.

Step 3: stop if the matrix has the desired number of groups; else, repeat step 2.

For an illustration, see Example 2.5.

2.3.5 Estimating missing or unavailable data

Missing data in the middle of a data stream may be calculated by using interpolation methods. Linear interpolation is fairly straightforward. For instance, if the demand values for September and November are

EXAMPLE 2.4

A company uses weight, density, length and width as attributes to define a product. The attribute data for four products on a numerical scale of 0 to 10 are given below. Compute the absolute distance metrics between the parts.

Product	*Weight*	*Density*	*Length*	*Width*
1	4	2	6	7
2	7	5	11	6
3	5	3	7	6
4	8	6	12	8

Let weight, density, length and width be attributes 1, 2, 3 and 4, respectively. The absolute Minkowski metric between products 1 and 2 is calculated as follows:

$$
\begin{aligned}
d_{12} &= |X_{11} - X_{21}| + |X_{12} - X_{22}| + |X_{13} - X_{23}| + |X_{14} - X_{24}| \\
&= |4 - 7| + |2 - 5| + |6 - 11| + |7 - 6| \\
&= 3 + 3 + 5 + 1 = 12
\end{aligned}
$$

Similarly, the distance metrics between all other products are calculated and summarized in the matrix below. Note that the matrix is symmetric and hence it is sufficient to consider the upper or lower triangular matrix. The matrix shows the dissimilarity between the various products. The smaller the distance between two products the larger the similarity. To convert this matrix to a similarity matrix simply compute the reciprocals of the distance metrics.

Product	*Product*			
	1	*2*	*3*	*4*
1	0	12	4	15
2	12	0	8	5
3	4	8	0	13
4	15	5	13	0

200 and 240, respectively, then October demand can be estimated as 220. The functional form of the data stream is required to interpolate missing values non-linearly. This may be done by using non-linear regression models.

Unavailable data usually relates to future information. Generally, this is estimated by using forecasting methods. The objective is not to deal

EXAMPLE 2.5

Partition the five products with the following similarity values into two groups using single linkage clustering.

Product	*Product*				
	1	*2*	*3*	*4*	*5*
1	–	0.2	0.5	1	0.8
2	0.2	–	0.4	0.7	0.6
3	0.5	0.4	–	0.3	0.9
4	1	0.7	0.3	–	0.5
5	0.8	0.6	0.9	0.5	–

Step 1: the similarity matrix is already given. Assume each product as a group. The desired number of final groups is two.

Step 2: the maximum value in the matrix is 1 and it corresponds to the similarity between products 1 and 4. Combine 1 and 4 into a new group 14. Compute the similarity between groups 14 and the rest of the groups. Similarity(14, 2) = min{similarity(1, 2), similarity(4, 2)} = min{0.2, 0.7} = 0.2. Similarly, compute other new similarity values. Drop groups 1 and 4. The new matrix is given below.

Group	*Group*			
	14	*2*	*3*	*5*
14	–	0.2	0.3	0.5
2	0.2	–	0.4	0.6
3	0.3	0.4	–	0.9
5	0.5	0.6	0.9	–

Since the number of groups is not equal to the desired number of groups, step 2 is repeated. The maximum value in the matrix is 0.9, corresponding to groups 3 and 5. Combine them into group 35. Recompute the similarity values. Drop groups 3 and 5 and insert group 35. The new similarity matrix is given below.

Group	*Group*		
	14	*2*	*35*
14	–	0.7	0.3
2	0.7	–	0.4
35	0.3	0.4	–

Repetition of step 2 yields a new group, 142, and we stop the procedure since the total number of groups is two. The final groups are {1, 4, 2} and {3, 5}.

with forecasting methods in this book. Several good textbooks are available in forecasting (Makridakis *et al.*, 1984) and also most books on operations management, inventory control and logistics present a chapter on forecasting. In this section, we present an introduction to regression-based methods for forecasting and a method known as 'booking profile forecasting' that is commonly used in the airline industry.

Regression models

Regression models identify a function that relates forecast to variables that drive the forecast. Forecast is the dependent variable and the drivers or the causal variables are the independent variables. Historical data on the variable to be forecasted and the drivers is used to identify the type and the parameters of the regression model. If the number of independent variables is one then it is called a simple regression; if there are more than one then it is referred to as a multiple regression. For instance, the demand for engines may depend upon the number of automobiles (simple). The demand for a particular brand of automobiles may depend upon employment rate, industrial growth rate, price, quality, competition and advertising. The relationship between the independent variable(s) and the dependent variable may be linear or non-linear.

The first step in developing regression models is to identify the causal variable(s). This is done based on experience, judgment and opinions from experts. The next step is to collect historical data on the dependent and independent variables. The third step is to plot and identify the functional relationship – linear, exponential or polynomial. The last step is to determine the parameters of the regression model. Table 2.1 provides a summary of the functional forms of a few regression models.

The exponential and the non-linear regression models may be transformed to a linear model by taking logarithms on both sides.

Table 2.1 Functional forms of regression models

Form	*Model*	*Dependent variable*	*Independent variable*	*Parameters*
Simple linear	$Y = aX + b$	Y	X	a, b
Multiple linear	$Y = aX_1 + bX_2 + cX_3 + d$	Y	X_1, X_2, X_3	a, b, c, d
Exponential	$Y = \exp(a + bX)$	Y	X	a, b
Non-linear	$Y = ab^X$	Y	X	a, b

Methods to estimate the values of the parameters are available in any standard statistics textbook (Walpole and Myers, 1978). Statistical software packages (SAS, BMDP, SPSS) are also available to perform regression analyses efficiently. Most packages can handle linear as well as non-linear regression models. An important measure of the goodness of the model is the correlation coefficient which is an indicator of the strength of the association between the dependent and independent variables. A value of 1 indicates a very high positive correlation and a −1 indicates perfect negative correlation. Examples 2.6 and 2.7 provide illustrations.

Booking profile forecasting

This method is useful for short-term forecasting of demand in situations where booking or ordering takes place a few days or weeks before the actual demand. It is widely used in the transportation industry, in particular in the airline industry (Hendricks and Kasilingam, 1993). It can be easily used where booking or ordering information is available.

EXAMPLE 2.6

Develop a simple regression model to estimate the customer satisfaction rating of a company given the following ratings over the past seven months:

Month	1	2	3	4	5	6	7
Rating	1.76	2.12	2.35	2.8	3.2	3.75	3.8

Let Y represent the rating, X represent the month and j be the index of the data. Let the regression model be $Y = a + bX$. The formulae for computing the parameters a and b and the correlation coefficient (γ) are as follows.

$$b = \frac{n \sum X_j Y_j - \sum X_j \sum Y_j}{n \sum Y_j - (\sum Y_j)^2}$$

$$a = \text{Xbar} - b\text{Ybar (Xbar is the average of } X \text{ values)}$$

$$\gamma = \frac{[n \sum X_j Y_j - \sum X_j \sum Y_j]}{[n \sum Y_j - (\sum Y_j)^2]^{1/2}[n \sum X_j - (\sum X_j)^2]^{1/2}}$$

Using equations (2.5)–(2.7), we compute the values of a and b as 1.3641 and 0.3654, resulting in the following regression model with a very high positive relationship between X and Y:

$$Y = 1.3641 + 0.3654X$$

EXAMPLE 2.7

Develop a methodology for fitting an exponential regression model for the satisfaction rating data given in Example 2.6. The exponential model is given as $Y = \exp(a + bX)$.

Taking logarithms on both sides yields $\ln Y = (a + bX)$. Letting $Z = \ln Y$ gives us the linear regression model, $Z = (a + bX)$. The parameters a and b can be computed (using Z and X) as in Example 2.6.

The method uses historical booking data and current booking information. There are two types of booking profile methods: the proportion method and the incremental method. In both cases, the first step is to construct booking profiles. This is done by aggregating data from several related bookings. For instance, booking data for the past several weeks may be averaged after eliminating the outliers. If the booking or ordering behavior depends upon the day of week of demand, then to get the booking profile for Thursday historical booking data for Thursday demand should be averaged. The proportion profile shows the percentage of demand booked by a certain day before the actual day of demand. In the airline terminology this is known as 'reading day'. Reading day 3 means 3 days before flight departure. The incremental profile shows the additional bookings or orders that will be placed between a given reading day and the departure day. The second step is to combine the booking profile and current booking information to obtain the forecast.

Two simple illustrations are presented to illustrate the booking profile construction process (Example 2.8) and the booking profile forecasting process (Example 2.9).

2.3.6 Probability distributions: form and parameters

Probability distributions are used to represent the uncertainty associated with the inputs to the logistics system under consideration or analysis. The key data required for determining the number and location of the receiving and shipping docks such as the arrival time of the trucks, truck loading and unloading times may not be known deterministically. In such cases, distributions are developed to capture the variability in the process. There are two important issues in assigning a distribution to a data stream. The first is the form (normal, Poisson or uniform) and the second is the values of the parameters (mean, variance, shape and size parameters). Again, several good statistical textbooks cover these two issues in great depth (Walpole and Myers, 1978).

EXAMPLE 2.8

The freight booking data (in pounds weight) for flight 732, London to Frankfurt, corresponding to 13th January 1997 and 20th January 1997 is given below [for the flight departing on 13th January 1997 by reading day 3 (10th January 1997) 630 lb were booked].

Departure date	*Reading day*						
	0	*1*	*2*	*3*	*4*	*5*	*6*
13 January 1997	1900	1330	930	630	430	230	150
20 January 1997	2000	1400	1000	600	500	300	200

Develop the proportion and incremental profiles for this flight. It is to be noted that both data points correspond to a Monday.

Proportion booking profile

First, the booking data are converted to cumulative percentage values. The cumulative percentage booking data for the two days is as follows [for example, for 20th January 1997 50% (1000/2000) is booked by reading day 2].

Departure date	*Reading day*						
	0	*1*	*2*	*3*	*4*	*5*	*6*
13 January	100	70	49	33	23	12	8
20 January	100	70	50	30	25	15	10

Assuming that both are representative data points (not outliers), the average of these two data points gives the following proportion booking profile.

Reading day	0	1	2	3	4	5	6
Average	100	70	49.5	31.5	24	13.5	9

Incremental booking profile

First, the booking data are converted to incremental booking information as follows. Incremental booking refers to the additional booking between a given reading day and the departure day. For instance, for 20th January 1997, additional booking between reading day 3 and departure day is 1400 (2000 – 600).

Departure date	*Reading day*						
	0	*1*	*2*	*3*	*4*	*5*	*6*
13 January 1997	0	570	970	1270	1470	1670	1750
20 January 1997	0	600	1000	1400	1500	1700	1800

Assuming that both are representative data points (that they are not outliers), the average of these two data points gives the following incremental booking profile.

Reading day	0	1	2	3	4	5	6
Average	0	585	985	1335	1485	1685	1775

EXAMPLE 2.9

On 24th January 1997 we are interested in forecasting the demand for flight 732, London to Frankfurt, departing on 27th January 1997. The current booking data for the departure of flight 732, London to Frankfurt, on that date is as follows.

Booking by:	*Amount booked*
21 January 1997	180
22 January 1997	270
23 January 1997	480

Using the profiles developed in Example 2.8 forecast the demand for flight 732 on 27th January 1997 (Note: the profiles developed in Example 2.8 can be used to forecast the demand as 27th January 1997 also is a Monday).

Proportion based forecast

The profile indicates that historically 24% is booked by reading day 4. Current booking data show an actual booking of 480 lb by reading day 4. Hence the expected demand by departure day is $(100 \div 24) \times 480\,\text{lb} = 2000\,\text{lb}$.

Incremental profile based forecast

Historically, the additional booking expected between reading day 4 and departure day is 1485 lb. Current booking data show an actual booking of 480 lb by reading day 4. Hence the expected demand by departure day is $(480+1485)\,\text{lb} = 1965\,\text{lb}$.

2.4 PHASE 3: PROBLEM ANALYSIS

This phase involves analysing the problem in hand in the light of the available data or the data that can be collected. The objective of the analysis is to develop alternative designs and select the best design or alternative with respect to the desired criteria. The type of analysis may range from basic spreadsheet calculations to sophisticated optimization modeling. The tools used for analysis may range from a simple spreadsheet to sophisticated operations research models. Some of the analysis may be performed using available software packages; others may require the development of new models, algorithms and software. In this section, a high-level description of some of the tools available for modeling and solving logistics problems is presented.

2.4.1 Mathematical models

Mathematical models refer to the general collection of linear, integer and non-linear programming models. These models typically have an objective function subject to a set of constraints. For instance, for a warehouse location problem a model may attempt to minimize the total distance traveled subject to constraints on the available number of locations, the budget and the maximum number of warehouses that can be located. The decision variables in a mathematical model may be continuous variables (gallons of fuel to be stored), general integers (number of automobiles to be shipped) or binary integers (whether to locate a warehouse or not). The model may be non-linear in the objective function and/or in one or more constraints. Simple examples to understand the formulation of mathematical models for production and logistics problems are discussed by Winston (1987). An array of mathematical models in production and inventory management is presented by Johnson and Montgomery (1974).

An analysis using mathematical models involves modeling and solving. In some cases, the model may be solved optimally by using efficient algorithms. In other cases it may have to be solved by using some approximate procedures. Regardless, the solution to a mathematical model is optimum or near-optimum only for the model – not for the real problem – since a model is only an abstract representation of reality. The effectiveness of a solution procedure is measured by using two attributes – speed and accuracy. Speed refers to the time it takes to solve a model on a computer. Accuracy refers to the closeness of the solution to optimality. Other effectiveness attributes include simplicity and ease of understanding and use of the model. A wide variety of mathematical models that are useful for various logistics decisions such as vendor selection, inventory control, warehouse location and transportation planning are presented in Chapters 3 through 8.

2.4.2 Heuristic methods

Heuristic methods are most commonly used in the areas of transportation planning and location modeling. They are extremely useful when optimal algorithms cannot solve the problem or take a very long time to solve because of problem size and complexity. Heuristic methods are analytical solution methods that are based on the intuitive understanding of the problem or the model that represents the problem. In general, a good understanding of the relationship between the decision variables and the various constraints and the objective function is required to formulate good heuristic procedures. There are also some standard approaches to developing heuristic methods. These can be classified as construction methods, improvement methods and construction and improvement methods.

In construction methods, as the name implies, a solution is developed from scratch by using certain rules that gear the development toward maximizing or minimizing the objective and simultaneously ensuring that the constraints are respected. Violation of constraints is checked at every step of the process. A good example is the development of a plant layout given the flow between the machines, size of the machines and the physical configuration of the building (Francis and White, 1974). A simple construction heuristic will start with placing two machines (say 1 and 2) with the greatest flow close to each other. Then, the next machine that has a high flow with the first two machines is placed close to machines 1 and 2. The procedure at this time checks to see if the physical configuration allows the machine to be placed there. Other process compatibility constraints are also checked. Sometimes, owing to the nature of the processes, certain type of machines cannot be placed close to each other even if the flow is very high. The procedure is continued until all machines have been placed.

Improvement methods start with a feasible initial solution and then make changes to the initial solution to seek improvement. The initial solution may be randomly generated or provided by the user. It may also be an output from a construction method. Any change that will result in an improvement from its current state will be accepted. In some cases, the improvement may have to be higher than an acceptable threshold level. The procedure stops when changes do not result in any additional improvement. The final solution in an improvement method depends upon three factors. First, the initial solution may have an impact on the changes that can be made and hence on the quality of the final solution. Second, the type of changes and perturbations that are done at every stage are relevant; for instance, if there are ten decision variables in the problem, at every state do we perturb only one variable or two at a time, or three at a time and so on? Last, the criteria for accepting the move at any given state has an impact on the final solution. Sometimes, it may be worthwhile to make a move that results in a negative improvement hoping that in subsequent steps a larger improvement may occur. All of the three factors mentioned above depend upon the type of problem in terms of its mathematical attributes of the objective function and constraints (convexity or concavity of the objective function and the characteristics of the constraint matrix).

In the construction and improvement methods, essentially the initial feasible solution is obtained by using a construction method. Again for the plant layout example, the solution from the construction method may be improved by exchanging the locations for any two machines. The method will exchange the locations for those two machines that will result in maximum improvement. An important consideration in designing heuristics is the ability to determine the goodness of the resulting final solutions and the time required to get the final solution. It is to be noted that some of the heuristics may provide the optimal solution to a problem. Whereas others may yield optimal solutions to certain instances of a problem.

2.4.3 Simulation methods

The methods described above are useful to solve problems analytically. However, because of complexity, the stochastic nature of data, the stochastic relationship between variables and certain problem-specific characteristics, not all real-world problems can be modeled and solved with use of mathematical models and/or heuristic methods. The use of mathematical or heuristic methods for such complex real-world problems will often require simplifying assumptions. This may cause the resulting solutions to be much inferior and sometimes infeasible. Simulation is probably the only available form of solution methodology in such cases. In a simulation model the complexities of the problem and the relationship between various variables are modeled by means of equations. The inputs to the model are often described by probability distributions. The system is then simulated over time to analyse the outputs. Simulation is typically used for analysing the performance of real-world systems over time under various operating procedures and policies. A carefully designed simulation model can also be used as a prescriptive tool for making decisions. In several real-world situations, decisions or alternatives from mathematical or heuristic methods are often tested by using simulation models to arrive at the best final decision.

2.4.4 Guidelines for strategy formulation

In some cases, a set of guidelines may help to address the issues or problems. However, guidelines are generally more appropriate for formulating high-level strategies. Analytical tools are needed to translate these strategies into planning or operational procedures. Some of the guidelines for developing logistics strategies are discussed in the following section (Ballou, 1992).

Total cost approach

One of the fundamental principles in the design and analysis of a logistics system is to consider all the relevant costs. Typical cost elements include production or purchase cost, inventory carrying cost, shortage cost, ordering cost, in-transit inventory carrying cost, storage cost, handling cost and transportation cost. Most of these costs are of a conflicting nature. Hence, any model or procedure developed for the analysis of the system must seek a trade-off between these costs. It is also important to understand the functional form of the costs to reflect discounts, one time increase in costs etc. In order to include the relevant costs, the scope and complexity of the system needs to be defined clearly. The scope must define the components of the system such as buyers' plants and warehouses, customers' stores and warehouses, vendors' locations and other third-party facilities. It must also define the key players involved in the functioning of the system such as buyers, sellers, internal employees, outside contractors and government and other private agencies.

Differentiated distribution

The basis of differentiated distribution is that not all products require the same level of customer service. Product characteristics, customer service needs and sales levels are different for different products. For instance, high-volume, high-value orders from customers should be served directly from plants and others may be served through warehouses. Fast-moving items are generally stored in warehouses that are near the customer locations and inside a warehouse near the pick-up points. Similarly, the amount of inventory and safety stock of an item depends upon the importance rating of the customer who is served by it. These differentiated strategies must be included as constraints or rules while developing a model or a procedure to analyse the system.

Postponement

The underlying philosophy behind postponement is to delay the manufacture and/or delivery of products until demand occurs. The objective is to reduce inventory-related costs. Postponement may be done in terms of time or product form. Under time postponement, shipping of items or goods to a location is delayed until demand occurs rather than shipping in anticipation. Under product form postponement, the final finishing of products such as painting, assembly, packaging and labeling are delayed until demand occurs. Examples for this include mixing of paints from base colors, storing products separately and packing them in 6 or 12 packs or stocking pipes in production sizes and cutting them to the required length based on demand.

Consolidation

The objective of consolidation is to realize the benefits of economies of scale. Consolidation may be done in time, location and size. An example of time consolidation is to deliver to two customers at the same location on Sunday when one customer needs the demand to be filled on Sunday and the other on Monday. It should be remembered that time consolidation may not be an option when high customer service levels are very critical. An example of location consolidation is the reduction of costs by shipping items to customers in nearby locations together. Matching delivery and pick-up from a customer at the same time when delivering and picking up from several local customers is an example of time and location consolidation. Hub and spoke operations of airlines and less-than-truckload carriers is an example of size consolidation. Smaller airplanes bring in passengers from smaller cities into a hub; a larger airplane carries them to another hub that is close to the destinations; then, smaller airplanes carry the passengers to the destinations. Sometimes, passengers to a common destination are served by a larger airplane directly from the first hub.

The strategies to be used such as postponement and consolidation must be reflected in defining the logistics problem. Specifically, the operating rules or constraints and the criteria for the selection of the best final system design are dependent upon these strategies.

2.5 PHASE 4: SYSTEM IMPLEMENTATION

The final phase in the analysis of a logistics system is the actual implementation of the selected alternative. Accurate and timely implementation is critical for the success of any new logistics system. Involvement of the users and buy-in from the management and the users are absolutely essential. Sometimes, the best alternative may have to be traded for the second best alternative in order to obtain the support and involvement from all concerned. There are two key steps to a successful implementation process. The first one is validation of the system and sensitivity analysis of the system performance to inputs. The second is user training and testing and actual turning over of the system to the users. Some of the key areas of focus in the final phase are documentation of the system, preparation of reference and user manuals and provision of technical and project support to the users.

The validation of the proposed alternative system design involves verifying if the outputs from the system are intuitive. Expert opinions may also be sought to understand the outputs. Results from the system analysis may also be compared with other similar systems. Sensitivity analysis is essential to understand the robustness of the proposed system. The outputs from the system should not vary too much for very small changes to the inputs. This is required since in real-life it is only possible to know the values of the inputs fairly accurately and not exactly. A good experimental design is vital for performing the sensitivity analysis.

User training is done in several different formats. The most commonly used format is to provide classroom training with demonstration followed by hands-on workshops. The users are also given reference and users' manuals during the training process. The testing process is generally done with use of a prototype model of the proposed system. The users use a prototype of the new system for a reasonable length of time. The purpose is to understand the new system, identify any problems with the system and to get comfortable with the system. During the process, the users tend to compare the performance of the system in terms of output quality, ease of use and time requirements with the existing process. After the user testing period the final production system is turned over to the users, with continuing technical support.

2.6 SUMMARY

In this chapter, the situations that warrant the design of a new logistics system have been discussed. The type of logistics system depends to a large extent on the attributes and the characteristics of the logistics product. The major phases of the logistics system analysis process was also outlined. Successful development of a good logistics system depends upon the entire process of logistics system analysis. All the phases of the analysis are equally important and critical. Some of the critical areas in each of the phases were also identified. The analysis process outlined in this chapter is one approach to analyse and design a logistics system. There are other possible approaches to this as well. However, it should be noted that all approaches are very similar and focus on the same basic steps but in different ways.

PROBLEMS

2.1 Give a few examples of logistics system design.

2.2 Discuss some of the logistics strategies and their relationships to business objectives.

2.3 Discuss the various dimensions used in defining a logistics product and its impact on logistics systems design.

2.4 A foundry production unit manufactures and sells 10 items. It is planning to develop a distribution system to satisfy the requirements of the 'A' items: items that represent no less than 70% of revenue. The annual revenue data (in thousands of dollars) of the 10 items for 1996 are as follows.

Items	1	2	3	4	5	6	7	8	9	10
Revenue	2.8	21.45	0.4	5	0.28	16.8	1.28	4.8	0.7	0.45

Determine the A items for the foundry unit.

2.5 Develop a methodology for fitting a non-linear regression model ($Y = ab^X$) for the satisfaction rating data given in Example 2.6.

2.6 An analysis of flight 700, Dallas to Frankfurt, departing on a Tuesday, indicates that booking starts as early as 8 days before departure. The historical booking profile based on proportions is as follows.

Reading day	0	1	2	3	4	5	6	7	8
Proportion	100	90	70	40	30	20	15	10	5

By 23rd January 1997 the airline's central booking office has received the following bookings for flight 700 departing on 28th January 1997:

Booking 1 from shipper A: 250 lb;
Booking 2 from agent 1: 500 lb;
Booking 3 from forwarder B: 1000 lb;
Booking 4 from John Ross: 250 lb.

What is the expected demand at flight departure?

2.7 In problem 2.6, what will be the forecast of expected demand at departure based on the following incremental booking profile?

Reading day	0	1	2	3	4	5	6	7	8
Incremental	0	2000	6000	12 000	14 000	16 000	17 000	18 000	19 000

2.8 XYZ Company is interested in clustering its current product offerings into two groups based on their weight, volume and value. The following matrix shows the similarity indices between the five different products.

Product	*Product*				
	1	*2*	*3*	*4*	*5*
1	1	0.8	0.2	0.7	0.3
2		1	0.2	0.6	0.3
3			1	0.4	0.8
4				1	0.3
5					1

Develop the product clusters using single linkage clustering.

2.9 Present a formulation of a suitable mathematical model for cluster analysis.

REFERENCES

Ballou, R.H. (1992) *Business Logistics Management*, Prentice-Hall, Englewood Cliffs, NJ.

Francis, R.L. and White, J.A. (1974) *Facility Layout and Location: An Analytical Approach*, Prentice-Hall, Englewood Cliffs, NJ.

Hendricks, G.L. and Kasilingam, R.G. (1993) Cargo revenue management at American Airlines. *AGIFORS Cargo Study Group Meeting*, Rome; copy available from author.

Johnson, L.A. and Montgomery, D.C. (1974) *Operations Research in Production Planning and Inventory Control*, John Wiley, New York.

Makridakis, S., Wheelwright, S.C. and McGee, V.E. (1984) *Forecasting: Methods and Applications*, John Wiley, New York.

Singh, N. and Rajamani, D. (1996) *Cellular Manufacturing Systems: Design, Planning, and Control*, Chapman & Hall, London.

Walpole, R.E. and Myers, R.H. (1978) *Probability and Statistics for Engineers and Scientists*, Macmillan, New York.

Winston, W.L. (1987) *Operations Research: Applications and Algorithms*, PWS-KENT Publishing, Boston, MA.

FURTHER READING

Armour, G.C. and Buffa, E.S. (1963) A heuristic algorithm and simulation approach to relative location of facilities. *Management Science*, 294–309.

Council of Logistics Management (1994) *Bibliography of Logistics Training Aids*, 2803 Butterfield Road, Suite 380, Oak Brook, IL 60521.

Kusiak, A. (1985) The part families problem in flexible manufacturing systems. *Annals of Operations Research*, 279–300.

Law, A.M. and Kelton, W.D. (1982) *Simulation Modeling and Analysis*, McGraw-Hill, New York,

Lee, R.C. and Moore, J.M. (1967) CORELAP – Computerized relationship layout planning. *Journal of Industrial Engineering*, 195–200.

Romesburg, H.C. (1984) *Cluster Analysis for Researchers*, Lifetime Learning Publications, Belmont, CA.

Shafer, S.M. and Rogers, D.F. (1993) Similarity and distance measures in cellular manufacturing, Part 1: a survey. *International Journal of Production Research*, 1133–42.

TRAINING AIDS: SOFTWARE AND VIDEOTAPES

Chicago Consulting, tel. (312) 346-5080: *SATISFY – Customer Service Software.*
Penn State Audio-Visual Services, tel. (800) 826-0312: *Business Logistics Management Series: Logistics Systems Analysis.*

CHAPTER 3

Logistics network planning

3.1 INTRODUCTION

As discussed in Chapter 1, logistics encompasses several functions such as vendor selection, transportation, inventory planning, warehousing and facilities planning and location. In addition, it is also impacted by production, marketing and product design decisions. Owing to the many interrelated functions and interfaces it is difficult to understand and measure the true logistics costs and to design effective and efficient logistics systems (Goetschalckx *et al.*, 1995). The requirements or demands placed on the logistics function have been changing over the past several years. These changing requirements may be grouped into the following three categories: competitive pressures, deregulation of transportation industry and information technology:

- Fierce competition and changing customer requirements dictate shorter product life-cycles, a larger variety of products and smaller manufacturing lot sizes. This results in a continuously changing product mix and a higher inventory of a larger variety of items, pushing up manufacturing and inventory related costs. To maintain a lower inventory and still provide a high customer service level one may adopt just-in-time manufacturing and distribution, resulting in the frequent movement of items in smaller quantities.
- The deregulation of the transportation industry has created several new transportation alternatives. This has resulted in innovations in transportation services, warehousing and storage options and in increased competition. The deregulation coupled with the increasing globalization of industries is leading to radically different logistics strategies.
- The entire field of information technology has been experiencing a phenomenal growth in terms of both hardware and software. Computers are available to handle large amounts of data, software packages are available to solve large and complex logistics planning and design problems and information and data can be exchanged and transmitted in real time through Electronic Data Interchange

(EDI). Identification of shipments and electronic transfer of shipment data can be done by bar codes and radio frequency tags. The locations of shipments and vehicles can be monitored by satellite.

3.2 LOGISTICS NETWORK MODEL

The changing requirements and environment have created the need for logistics as a competitive tool to improve customer service and reduce the total cost of providing customer service. The changes also have created the need constantly to review and redesign the various logistics systems and tools used by companies. The design of a logistics system is based on four major planning areas: customer service levels, location decisions, inventory planning and transportation management. Customer service in logistics includes product availability, lead time to obtain the product, condition of the product when received and accuracy of filling an order. Location decisions relate to the placement of facilities such as warehouses, terminals, stores and plants and the assignment of demands to supply points. Inventory planning encompasses setting up inventory levels and inventory replenishment schemes. Transportation management deals with transportation mode, fleet size, route selection, vehicle scheduling and freight consolidation. All four areas are economically interrelated and should be planned in an integrated manner to achieve maximum benefit. Methodologies and systems that deal with integrated planning typically are at an aggregate level and do not include detailed problem definition. Systems and procedures that are more detailed do not address all four areas simultaneously. The primary reason is size and complexity.

Logistics network modeling tools attempt to include as much detail as possible but still address the logistics system design problem in an integrated manner. Some of the questions answered by an integrated logistics network model are as follows:

- the number of warehouses, their location, ownership (private or public) and their size;
- the allocation of customer demand to supply points (warehouses or plants); allocation to single or multiple supply points;
- the amount of inventory to be maintained at various locations;
- the type of transportation services to use;
- the level of customer service to be provided (that can be provided).

Some of the complexities involved in designing a logistics system with use of network modeling are as follows:

- the integration of vehicle routing and scheduling;
- the uncertainty in demand, which requires demand to be forecasted;

- the identification and development of the appropriate type of cost functions;
- the dynamic nature of the demand and cost functions over a period of time;
- dependency relationships between inventory and transportation decisions;
- the relationship between customer service levels and key logistics decisions;
- the size of the problem.

Determination of the optimal logistics network configuration is a fairly complex task because of the large number of vendors and customers, the hundreds of candidate locations for warehouses and plants and the extremely large number of transportation options. Despite these complexities and challenges, several large corporations periodically analyse their distribution network in an effort to reduce costs and/or improve customer service. Studies have indicated that improvements to logistics networks have typically resulted in 5% to 15% savings in logistics costs.

3.3 LOGISTICS NETWORK MODELING APPROACH

Most logistics network design and planning approaches fall into two categories. One approach is to use very flexible and generic models which are solved by using commercially available mixed integer programming software packages. The other approach is to develop highly specialized models with specialized solution procedures to arrive at the design. Under the second approach the model is specific to a given application; hence it represents the design problem in hand more accurately. Also, efficient solution procedures may be developed to solve large-scale logistics network design problems in a reasonable amount of time. However, they require a significant amount of time and resources to build the model and the solution procedure. In contrast, generic models are easy to understand and to apply to a situation. However, they may not represent the problem completely and accurately. Also, the amount of computational time and memory may be excessive to solve industrial logistics design problems; sometimes it may even be impossible to solve the generic models beyond a certain problem size.

An ideal approach would be to have a logistics network model that is generic and comprehensive enough to represent the logistics system with sufficient accuracy but at the same time be amenable to specialized solution procedures in order to solve large size design problems within a reasonable amount of computer time. The model must consider the movement and storage of raw material or components from raw material sources to plants either directly or through warehouses and the movement and storage of finished products or components from plants to customers either directly or through distribution centers. The network model represents the various logistics activities as nodes and arcs. The nodes represent locations of facilities where product flows are created such as plants, suppliers, warehouses and customers. There may be several types of warehouses (public or private), vendors and plants with different capacities, cost structures and technologies. The arcs represent the flow of raw material, products or components between the facilities. The flows include several transportation options with varying transport characteristics and costs. Ideally, the model must be able to handle multiple facilities (vendors, warehouses, plants and customers), multiple commodities (raw material and finished goods) and multiple time periods. The trade-off between inventory and transportation costs and between customer service and transportation or inventory costs must be implicitly considered by the model.

3.4 COMPONENTS OF THE LOGISTICS NETWORK MODEL

The basic components of a logistics network are vendors, products, plants, warehouses or distribution centers, transportation services and customers. A brief discussion of each of the components is given in this section.

3.4.1 Vendors

The potential vendors and their locations is one of the inputs to a network model. Associated with the vendors are the availability of various products, lead time, quality and price. The locations of the vendors also dictate the type of transportation services from vendors to plants. There may be limits on the maximum number of vendors to be used for a given product. Vendors may offer quantity discounts for certain products over a certain order size and a group discount if certain combinations of products are ordered.

3.4.2 Products

A set of products or a group of commodities that move through the supply chain is one of the inputs to the model. Each commodity has certain physical characteristics such as weight, size and volume. Also, each commodity has a certain value and demand. The unit of measurement may be truckload or car load, pallet load or unit load.

3.4.3 Plants

These are locations where products or commodities are manufactured, assembled or purchased from. Each plant has a location, the capability to produce certain products and production capacity restrictions. There are fixed costs for establishing the facility and variable costs for manufacturing a product. Also, finished products stored at the plant incur inventory costs.

3.4.4 Distribution centers

There may be several distribution centers between production facilities and demand centers as well as between vendor locations and production facilities. The distribution centers may also have a hierarchical structure. For instance, products may be shipped from plants to regional warehouses, then to smaller, local warehouses and finally to customers close to the local warehouses. All warehouses have limitations in terms of storage space, type of products that can be handled and throughput or handling capacity. The resources required to handle products at different warehouses may be different. The associated costs of building and operating warehouses are the initial cost of establishing a warehouse and the recurring costs to maintain and operate the warehouse. The number and size of warehouses and the amount of inventory to be held are some of the major warehouse planning and operational decisions.

3.4.5 Transportation services

These services include both local delivery and pick-up operations as well as over-the-road or trunking services. Transportation decisions include the mode of transportation, shipment size and allocation of product flow from source to sink to various transportation modes. Each transportation mode has restrictions in terms of capacity and availability. Other important characteristics to be considered include transit time, transit time variability, costs and the number of carriers. Costs include all fixed and variable transportation costs and in-transit inventory costs. Movement of products between two places, for instance from a plant to a warehouse, may be split among multiple transportation modes. Within the same mode, it may be split among different carriers. The transportation decisions made at the network level are aggregate in nature and hence will not concern tactical decisions such as vehicle routing and scheduling.

3.4.6 Customers

Customer service is measured in terms of product availability and the lead time required to obtain unavailable products. A customer may be served by only one warehouse or plant or by more than one plant or warehouse. Travel time and travel distance are indicators of customer service. Order processing time along with the actual transit time determines the lead time. Customer demand is one of the key inputs; demand must be known in terms of location, time of occurrence, amount and the type of product.

3.5 GENERALIZED LOGISTICS NETWORK MODEL

In this section, a verbal formulation of a generic logistics network model is presented. The objective function of a logistics network design model minimizes the fixed facility costs for warehouses, vendors and plants, the operating costs for plants and warehouses, transportation costs, local delivery costs, inventory costs and in-transit inventory costs. The objective function is optimized with respect to five major categories of constraints. The first one is related to capacities of plants, warehouses and transportation carriers. The second is related to the internal consistency of the model; for example, a warehouse can serve a customer only if it is built and a plant can supply to a warehouse only if it is operating. The third set of constraints express customer service requirements in terms of maximum distance between warehouse and customer, customer demand and minimum inventory levels or safety stock. It also includes the requirements of plants in terms of raw materials or components. The fourth set of constraints ensure conservation of flow between the various components of the logistics network. Finally, there are constraints on the decision variables in terms of their non-negativity and/or integrality. All inputs to the model should be based on the same time period – day or week.

The modeling of a logistics network requires a complete and accurate understanding of the various costs and the relationship between them. Some of the costs may be linear, some may be non-linear and others may be stepwise linear or non-linear. For example, production costs may be linear or non-linear up to a certain number of units, after which an additional fixed cost may have to be incurred. When most of the production costs are variable the cost per unit will be linear within a particular range. Non-linear transportation costs such as quadratic cost functions or exponential cost functions are more realistic (Jara Diaz, 1982) but are much more complex and difficult to model and solve. Hence non-linear cost functions are typically approximated by piecewise linear costs functions. The precise functional form of costs may be obtained by using historical data. For instance, regression analysis of historical warehouse cost data may indicate the following relationship [equation (3.1)] for warehouse operating cost. The first term on the right-hand side is the storage cost and the second term on the right-hand side is the handling cost (note that the warehouse operating cost is the sum of storage cost and handling cost to meet the demand at customer locations):

$$\begin{aligned}\text{warehouse operating cost} = {} & [20 \times (\text{storage rate}) \times (\text{demand})^{0.6}] \\ & + [(\text{handling rate}) \times (\text{demand})]\end{aligned} \tag{3.1}$$

3.5.1 Model specification

The generalized logistics network model may be formulated as follows.

$$\begin{aligned}\text{Minimize } Z = {} & \text{vendor establishment cost} + \text{plant fixed cost} \\ & + \text{purchase cost for components or raw materials} \\ & + \text{production cost} + \text{plant inventory cost} \\ & + \text{warehouse inventory cost} + \text{in-transit inventory cost} \\ & + \text{transportation cost} + \text{local pick-up or delivery cost} \\ & + \text{warehouse handling cost}\end{aligned} \tag{3.2}$$

Subject to

vendor supply to all plants $\leq$ vendor capacity
plant supply to all warehouses $\leq$ plant capacity
warehouse supply to customers $\leq$ warehouse capacity
transportation between points $\leq$ carrier capacity (weight and volume)
vendor, plant or warehouse can supply only if open
receipt from vendors $\geq$ plant requirements
receipt from plants or vendors $\geq$ warehouse requirements
receipt from warehouses, plants or vendors $\geq$ customer requirements
time for order processing and transportation $\leq$ acceptable lead time
demand exceeding inventory $\leq$ acceptable stock out level
conservation of flow at any point: inbound = outbound + usage
non-negativity and integrality constraints

3.5.2 Generic solution procedure

In most cases, the formulation of the model determines the type of solution procedure that can be employed, and the type of solution procedure determines the size of the network model that can be solved, computer memory requirements, time to find a solution and the ability to find an optimal solution (Ballou and Masters, 1993). The model presented above falls into the category of large-scale mixed integer programs. It is much more difficult to solve these models than linear or network flow programs of comparable size. Appropriate preprocessing methods may reduce some of the redundant constraints and variables. Preprocessing may also be used to aggregate or combine variables where feasible. In some cases, the model may be decomposed by plant, customer, vendor or warehouse to obtain subproblems and a master problem. In other cases, some of the constraints may be dualized and a procedure based on Lagrangian relaxation and subgradient optimization may be used. Heuristic procedures always come in handy as the last resort. For large logistics network planning models (most real-life problems are very large), optimality may never be achieved under any of the above methods. Heuristic procedures typically run fast and use less computer memory. However, with some heuristic methods it may not be possible to know how far from optimal the solution is.

3.5.3 General guidelines in using the network model

Formulating a model and solving it to develop the logistics network does not warrant any success. There are three major guidelines that need to be followed (Napolitano, 1997). The first guideline is to validate the model before performing any scenario analysis. One approach is to validate results based on history. Another approach is to validate against intuition and simple common sense. Sensitivity analysis of changes to input data is also necessary to test the robustness of the model. Any major conclusions or decisions must be made only after model validation. The second guideline is to remember that the solution from the logistics network model is not the final answer. Human expertise, judgment and common sense have to be used to modify the answer for implementation. The most common reasons are that an optimal solution is best with respect to a chosen model and almost all models fail to capture reality completely. When they do they are so large and complex that they have to be solved heuristically, resulting in suboptimal solutions. In general, it is a lot cheaper and quicker to use existing software packages for distribution modeling. The loss of exactness is more than compensated for by savings in cost and time. The last guideline is that logistics network modeling may not be the best approach to solve network design problems. In several cases the cost of modeling may outweigh the associated benefits. In several cases, simple rules of thumb will address location, size and assignment of warehouse to customer decisions, mainly because the number of candidate locations is limited or system constraints arising from contract or legal reasons limit the available transportation and storage options.

3.6 SUMMARY

An integrated approach to logistics network planning addresses the interrelationships and trade-offs between the various functions. Major decisions in all the functional areas are addressed simultaneously rather than hierarchically. This results in better overall decisions. Logistics network planning is a very complex and difficult problem to model and solve. Recent advances in operations research in the areas of mixed integer programming and network modeling enables us to solve large, real-life logistics network planning problems through optimal procedures. Also, several commercial software packages, both generic and specialized, are available to solve large problems efficiently. Some of the popular generic mixed integer programming software libraries to solve large problems include CPLEX and ILOG (details of software and software libraries are given at the end of this chapter). Examples of specialized logistics network design packages include the SAILS and CAPS Logistics tool kit. Almost all of these software packages are available in PC DOS/ Windows and UNIX versions. Logistics network models analytically represent the flow of products through the companies' supply chain. It is to be recognized that these models are merely tools. As with any other approach, good judgment and common sense are still needed to select the final design.

PROBLEMS

3.1 Show graphically the relationship between inventory cost and transportation cost. Assume that a lower inventory level may be compensated for by using a premium transportation option to provide the same level of customer service.

3.2 Sketch the various possible functional forms of the production cost of a product. Explain under what conditions they are valid.

3.3 Discuss the advantages and disadvantages of using an integrated logistics network planning methodology.

3.4 List the various decisions involved in designing a logistics network.

3.5 Represent graphically a logistics network with three vendors, two plants and two warehouses to supply products to six customers.

3.6 For problem 3.5, assuming a single product and linear cost functions, formulate a mixed integer programming model to represent the network. State your other assumptions clearly.

3.7 What are the effects of various logistics network components on cycle time?

3.8 Discuss the impact of different logistics network decisions on product stock out.

REFERENCES

Ballou, R.H. and Masters, J.M. (1993) Commercial software for locating warehouse and other facilities. *Journal of Business Logistics*, **14**, 71–105.

Goetschalckx, M., Cole, M.H., Dogan, K. and Wei, R. (1995) A generic model for the strategic design of production–distribution systems. Research Report, School of Industrial and Systems Engineering, Georgia Institute of Technology, Atlanta, GA.

Jara Diaz, S.R. (1982) The estimation of transport cost functions: a methodological review. *Transport Reviews*, **2**, 257–78.

Napolitano, M. (1997) Distribution network modeling. *Industrial Engineering Solutions*, **6**, 20–5

FURTHER READING

Aikens, C.H. (1985) Facility location models for distribution planning. *European Journal of Operational Research*, **22**, 263–79.

Cole, M.H. (1995) *Service Considerations and the Design of Strategic Distribution Systems*. Unpublished PhD dissertation, School of Industrial and Systems Engineering, Georgia Institute of Technology, Atlanta, GA.

Copacino, W. and Rosenfield, D.B. (1985) Analytical tools for strategic planning. *International Journal of Physical Distribution and Materials Management*, **15**, 47–61.

House, R.G. and Karrenbauer, J.J. (1982) Logistics system modeling. *International Journal of Physical Distribution and Materials Management*, **12**, 119–29.

Perl, J. and Sirisoponslip, S. (1988) Distribution networks: facility location, transportation, and inventory. *International Journal of Physical Distribution and Materials Management*, **18**, 18–26.

TRAINING AIDS: SOFTWARE AND SOFTWARE LIBRARIES

Software

ADLnet Software, A.D. Little, Inc., tel. (617) 864-5770.

CAPS Logistics Tool Kit, CAPS Logistics, Inc., tel. (404) 432-9955.

NETWORK, Department of Operations Research, Case Western Reserve University., tel. (216) 368-3808.

SAILS, Insight Inc., tel. (703) 683-3061.

Software libraries

CPLEX (1997) CPLEX Optimization Inc., Incline Village, NV, USA.

ILOG Optimization (1997) ILOG Ltd., Bracknell, Berks, UK.

CHAPTER 4

Vendor selection

4.1 INTRODUCTION

As early as the nineteenth century, purchasing, which is a division of logistics, was regarded as an independent and important function by many of the US railway organizations. The first book that was specifically devoted to purchasing, entitled *The Handling of Railway Supplies – Their Purchase and Disposition*, was published in 1887. Since the beginning of the twentieth century there were several movements in the evolution of purchasing or materials management functions, as depicted in Figure 4.1.

Prior to World War 2 the success of a firm did not depend very much on what it could sell since the market was unlimited. Instead, the ability to obtain the required materials and services from the vendors was the key to organizational success. Attention was given to the organization, policies and procedures of the purchasing function, and it emerged as a recognized managerial activity. During the 1950s and the 1960s purchasing continued to gain stature as the techniques for performing the function became more refined and as the supply of trained and competent professionals to make sound purchasing decisions increased. In the 1970s organizations faced two major problems: the international shortage of almost all basic materials needed to support operations and a rate of price increases far above the norm since the end of World War 2. These developments put the spotlight directly on the purchasing departments for their performance in obtaining the required materials from vendors at reasonable prices.

In the 1990s it became clear that organizations must have an efficient and effective purchasing and materials function if they were to compete successfully in the domestic and international markets. The future will

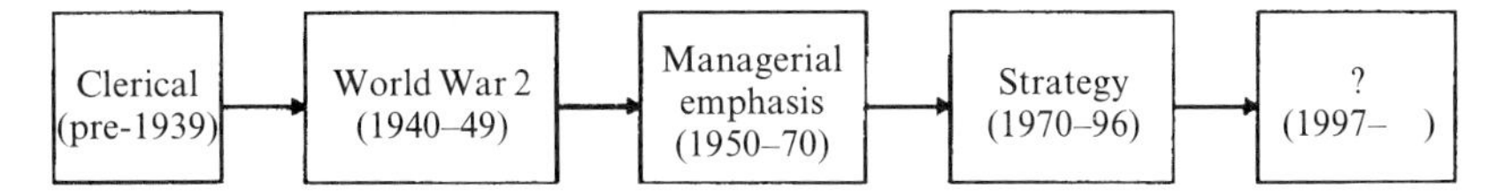

Figure 4.1 Evolution of materials management over time.

see a gradual shift from the predominantly defensive strategies to aggressive ones in order to remain competitive. Organizations will take an imaginative approach for achieving their materials management objectives to satisfy long-term and short-term goals. The field of purchasing, which is a subset of materials management, centers on supplier or vendor selection. Other activities that are also associated with purchasing are (Ballou, 1992):

- selecting and qualifying suppliers;
- rating supplier performance;
- negotiating contracts;
- timing of purchases;
- predicting price, service and demand changes;
- specifying mode of shipping.

One of the major challenges for today's purchasing managers in any industry, be it service or manufacturing, is selecting the right vendor(s) for their components, parts and supplies. Selection of vendors includes determining the vendors and the amount of order to be placed on each vendor. Choosing the appropriate vendor(s) is extremely important since total quality management and customer satisfaction are the goals of almost all the organizations. The performance of vendors has a significant impact on the productivity, quality and competitiveness of an organization. The importance and difficulty of selecting suppliers is complicated by the latest business trends, which include: the increase in the value of purchased parts as a percentage of total revenue for manufacturing firms; growth in imported parts and supplies; and the increased rate of technological change accompanied by shorter product life-cycles. For example, the cost of components and parts purchased from outside vendors is approximately 50% of sales for large automotive manufacturers. High-technology firms purchase materials and services up to 80% of their total product costs (Burton, 1988). Japan purchases up to 40% more supplies and materials for their automobile industries compared with their counterparts in the United States and Europe (Leenders and Fearon, 1993). Most organizations typically spend 40% to 60% of the revenue of their end products on purchased parts (US Bureau of the Census, 1983, pp. 5–8 and appendix).

Selection of vendors based on quality, price, delivery, service and capacity generally ensures buyer satisfaction. This is shown in Figure 4.2.

The initial purchase price of an item is only one element of the total cost. There are other associated costs such as cost of establishing vendors, transportation and storage costs and costs of receiving poor quality material. Companies try to achieve a balance between price and value of a purchased part or material during the acquisition process. Usually vendors are selected by their ability to offer the best cost or

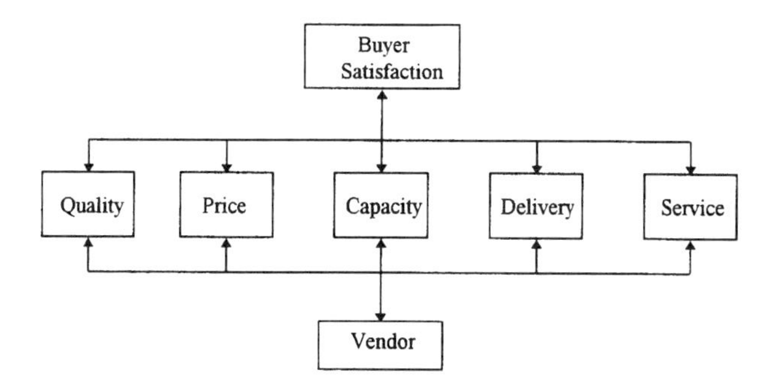

Figure 4.2 Desirable vendor criteria.

quality package. Quality level may be specified to the vendors in a variety of ways (Lee *et al.*, 1990): commercial standards, design specifications, samples, market grades, brand or trade names, functional specifications and tolerances. There are several other additional factors that need to be considered as well while selecting vendors. Past performance, facilities, technical expertise, financial status, reputation, organization and management, compliance to procedures, communications, labor relations and the location of the vendors are some of them. One of the pioneering research results on vendor selection by Dickson (1966) provides a benchmark on the trends in the importance of vendor selection. The article documented the multi-objective nature of vendor selection and ranked the importance placed on 23 selection criteria by 170 industrial purchasing agents and managers.

There are several methods available to select vendors and to determine the quantities to be ordered from the selected vendors. These methods can be broadly classified into descriptive, empirical and optimization based approaches. The descriptive methods select vendors by evaluating the qualitative factors related to the vendors such as reputation, expertise, organization and communications. The empirical methods evaluate vendor characteristics through relative weighting schemes. The optimization methods minimize various costs associated with purchasing to meet certain requirements in terms of quality, lead time and demand.

4.2 CLASSIFICATION OF VENDOR SELECTION DECISIONS

Soukup (1987) classifies vendor selection decisions into three major categories.

1. Potential vendors are similar under all foreseeable circumstances; only minor variations in performance can be expected. In this

situation the probability of selecting the wrong vendor is very high; however, the consequences of such an error are less significant.
2. The potential vendors differ significantly; one vendor is clearly superior under all foreseeable conditions. In this situation the consequences of making the wrong selection are very serious; however, the probability of making an error is very small.
3. The candidate vendors differ significantly; the best vendor under some circumstances may not be the best under other circumstances. This presents the purchaser with a relatively high probability of making the wrong decision as well as the possibility of serious consequences of the error.

For categories 1 and 2 the vendor selection decision can be handled as a routine decision. Category 3 calls for vendor selection with considerable effort and care. Table 4.1 summarizes the classification of vendor selection decisions.

The objective of this chapter is to present several approaches to address the vendor selection decisions that fall under category 3. First, a few empirical methods based on rating of vendors on various criteria are described. Next, an approach based on the analytical hierarchic process to determine the weights to be used in the empirical methods is presented. Finally, a few optimization based approaches that can be used to select more than one vendor simultaneously are presented.

4.3 EMPIRICAL METHODS

Empirical methods essentially use some kind of relative importance or weights for the various criteria and/or ratings or rankings of the vendors with respect to the selection criteria. The selection criteria may include objective as well as subjective factors.

Table 4.1 Classification of vendor selection decisions

Category	*Probability of error*	*Consequence of error*	*Decision type*
1. Vendors are similar under all conditions	High	Very small	Routine
2. Vendors differ significantly; one vendor is superior under all conditions	Low	High	Routine
3. Vendors differ significantly; the best vendor depends on certain future conditions	High	May be very high	Complex; requires careful analysis

Table 4.2 Summary of Dickson's (1966) findings

Rank	*Factor*	*Mean rating*	*Importance*
1	Quality	3.508	Extreme
2	Delivery	3.417	
3	Performance history	2.998	
4	Warranties and claim policy	2.849	
5	Production facilities and capacity	2.775	Considerable
6	Price	2.758	
7	Technical capability	2.545	
8	Financial position	2.514	
9	Procedural compliance	2.488	
10	Communication system	2.426	
11	Reputation and position in industry	2.412	
12	Desire for business	2.256	
13	Management and organization	2.216	
14	Operating controls	2.211	
15	Repair service	2.187	Average
16	Attitude	2.120	
17	Impression	2.054	
18	Packaging ability	2.009	
19	Labor relations record	2.003	
20	Geographical location	1.872	
21	Amount of past business	1.597	
22	Training aids	1.537	
23	Reciprocal arrangements	0.610	Slight

4.3.1 Factor analysis

The first step in factor analysis is to identify the important factors or criteria to be considered in selecting the vendors. The findings of the vendor selection survey by Dickson (1966) may help in this step. Table 4.2 provides a summary of Dickson's findings regarding the importance of 23 vendor selection criteria. The second step is to rate each vendor against all the factors on a scale of 1 to 10. The third step is to combine the ratings for all the factors for each vendor. The last step is to select the vendor with the highest combined score.

Factor analysis model

$$v(j) = \sum_{i=1}^{n} s(i, j) \tag{4.1}$$

where,

$v(j)$ is the combined rating for vendor j;
$s(i, j)$is the score for vendor j on factor i;
n is the number of factors.

EXAMPLE 4.1

The purchasing department of a company considers price, lead time and quality as the three important factors in selecting a vendor from among three vendors. The first vendor is rated as 6, 4 and 6 on the three factors, respectively. The second vendor is rated 5, 9 and 7, respectively, and the third vendor is rated 8, 5 and 7, respectively. The combined rating for vendor 1 is 16, for vendor 2 is 21 and for vendor 3 is 20. Hence, vendor 2 will be selected. It is to be noted that this method assumes equal importance for all the factors.

An illustration is provided in Example 4.1.

4.3.2 Weighted factor analysis

This is a modification of the factor analysis method where weights are assigned to the different criteria based on their importance (Wind and Robinson, 1968).

Step 1: identify the vendor selection factors or criteria and their weights (relative importance).
Step 2: rate each vendor on all the criteria.
Step 3: compute a weighted rating for each vendor and select the vendor with the highest weighted rating.

Weighted factor analysis model

$$v(j) = \sum_{i=1}^{n} w(i)s(i,j) \tag{4.2}$$

where,

$v(j)$ is the weighted rating for vendor j;
$w(i)$ is the weight for factor i;
$s(i,j)$ is the score for vendor j on factor i;
n is the number of factors.

An illustration is given in Example 4.2.

Weighted factor analysis is one of the most powerful vendor analysis methods. It is simple and easy to use and is relatively inexpensive to implement. The weakness of the method lies in the difficulty involved in estimating the weights and in quantifying some of the subjective factors. Also, the ratings of the vendors against the various criteria are deterministic or point estimates.

EXAMPLE 4.2

Suppose in Example 4.1 the weights for the three factors (price, lead time and quality) are 0.4, 0.2 and 0.4, respectively. The combined rating for vendor 1 will be equal to $(0.4 \times 6) + (0.2 \times 4) + (0.4 \times 6) = 5.6$. The combined ratings for the other two vendors are computed in a similar manner, yielding 6.6 and 7, respectively. Now vendor 3 will be selected. This is because of the greater importance placed on price and quality compared with lead time.

4.3.3 Vendor profile analysis

This method is a Monte Carlo simulation version of the weighted factor method (Thompson, 1990).

Step 1: define the range of possible values for each criteria and assume a distribution for the values. The assumption of uniform distribution for these values adequately describes vendor performance for most real-life situations.

Step 2: for each vendor, randomly select values for all the criteria and compute the combined rating by means of the weighted factor method. Repeat this step for several iterations (say $k = 500$, where k is the maximum number of iterations). Then a frequency distribution is created by using the combined ratings from k iterations.

Step 3: repeat step 2 for all the vendors. Now one has frequency distributions for all the vendors.

Step 4: the final step involves comparing the frequency distributions of the various vendors in terms of their mean (or most appropriately mode), variance and degree of overlap.

Obviously, one would like to select the vendor with a high mean and low variance. One should also select the vendor for whom the distribution has minimum overlap with the distributions of other vendors. Extensive overlap between the distributions of the vendors indicates that the vendors are really different from one another. An illustration is provided in Example 4.3.

4.3.4 Analytical hierarchic process

The analytical hierarchic process (AHP) (Narasimhan, 1983) is useful in estimating the weights to be used for the various criteria as well as in quantifying the subjective factors. AHP provides a systematic approach

EXAMPLE 4.3

Let us assume that the values for all three criteria mentioned in Example 4.2 follow uniform distribution with range [0, 10]. The maximum number of iterations is set to 100. For vendor 1, the random sampling for the first iteration yields 7 for the first criteria, 6 for the second and 8 for the third. The rating for vendor 1 for iteration 1 is then $(0.4 \times 7) + (0.2 \times 6) + (0.4 \times 9) = 7.6$. Similarly, ratings for vendor 1 may be computed for 100 iterations and a frequency distribution may be constructed. By continuing the Monte Carlo simulation one obtains the frequency distributions for the three vendors, as shown in the following figure.

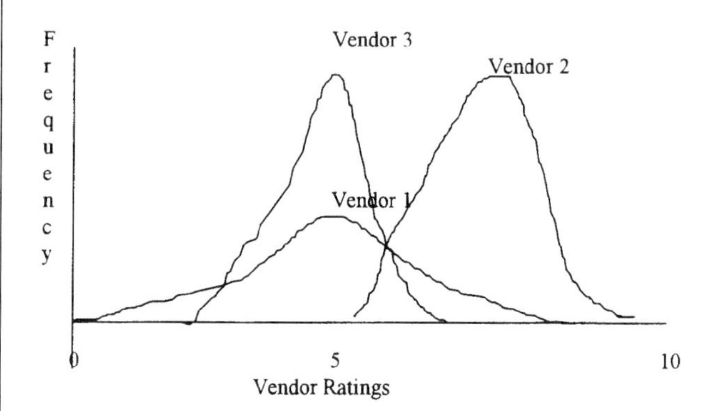

The figure indicates that the modal performance of vendor 2 is significantly higher than that for vendors 1 and 3. Though vendors 1 and 3 have similar performance in terms of mode, vendor 3 has a significantly lower variance, indicating that there is less risk in choosing vendor 3. Finally, the frequency distribution of vendor 2 has minimal overlap with the distributions of vendors 1 and 3, which indicates that the worst performance from vendor 2 approximately equals the best performance of the other vendors.

to select vendors, focusing on commonly used vendor selection criteria.

The first step involves formalizing the vendor selection into a hierarchical model. Figure 4.3 shows an example of the hierarchical model of the supplier selection problem, with selection criteria of price, delivery and quality. The criteria make up the components of the first level. The second level consists of the dimensions of the components represented in the first level. For instance, quality has 'fraction defective' and 'repair costs' as two of its dimensions. The lowest level shows the available vendors.

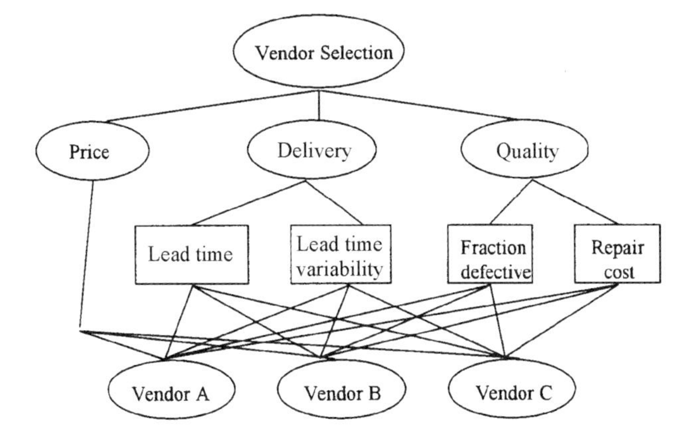

Figure 4.3 Hierarchical model for vendor selection.

The second step is to set the relative weights for the criteria at the first level, the relative weights for the dimensions of each criteria at the next level and the relative ranking of the vendors with respect to each dimension. This is done by performing pairwise comparisons. The scale used for pairwise comparisons conforms to the definitions given in Table 4.3.

The pairwise comparisons for step 2 are given in Tables 4.4–4.8. Table 4.4 shows the pairwise comparisons among the criteria (price, delivery and quality) for the first level. In Table 4.4, the first row shows

Table 4.3 Scale for pairwise comparison

Importance value	*Definition*
1	Equal importance
3	Weak importance
6	Strong importance
9	Absolute importance
Reciprocals of above non-zero values	If activity *j* has one of the above non-zero numbers assigned to it when compared with activity *k*, then *k* has the reciprocal value when compared with *j*

Table 4.4 Pairwise comparisons of criteria (level 1) illustrated in Figure 4.4

Criteria	*Price*	*Delivery*	*Quality*
Price	1	3	1
Delivery	$\frac{1}{3}$	1	$\frac{1}{3}$
Quality	1	3	1

Table 4.5 Pairwise comparisons of delivery dimensions illustrated in Figure 4.4

Dimension	*Dimension*	
	Lead time	*Lead time variability*
Lead time	1	5
Lead time variability	$\frac{1}{5}$	1

Table 4.6 Pairwise comparisons of quality dimensions illustrated in Figure 4.4

Dimension	*Dimension*	
	Fraction defective	*Repair cost*
Fraction defective	1	$\frac{1}{5}$
Repair cost	5	1

Table 4.7 Pairwise comparison of vendors on price illustrated in Figure 4.4

Vendor	*Vendor*		
	A	*B*	*C*
A	1	1	$\frac{1}{5}$
B	1	1	$\frac{1}{3}$
C	5	3	1

the pairwise comparisons of price with price and other components. The entries in the first row indicates that price is of weak importance to delivery and of equal importance to quality.

The third step is to compute the relative weights for the criteria and dimensions. A rigorous approach to compute the relative weights is based on an eigenvector method. An approximate procedure consists of computing the geometric averages of the row entries and normalizing the averages. This step is illustrated for the pairwise comparisons of the criteria given in Table 4.4. For instance, the geometric average of row 1 in Table 4.4 is 1.4422. The geometric averages of the other rows are 0.4808 and 1.4422, respectively. The normalized weights (which sum to 1) based on the geometric averages for price, delivery and quality are 0.43, 0.14 and 0.43, respectively. This step is illustrated in Table 4.9.

Steps 2 and 3 may be repeated for the remaining pairwise comparisons given in Tables 4.5–4.8. The resulting hierarchical model with

Table 4.8 Pairwise comparison of vendors on: (a) lead time; (b) lead time variability; (c) fraction defective; (d) repair cost, illustrated in Figure 4.4

Vendor	*Vendor*		
	A	*B*	*C*
(a)			
A	1	5	3
B	$\frac{1}{5}$	1	1
C	$\frac{1}{3}$	1	1
(b)			
A	1	$\frac{1}{3}$	2
B	3	1	5
C	$\frac{1}{2}$	$\frac{1}{5}$	1
(c)			
A	1	$\frac{1}{2}$	$\frac{1}{3}$
B	2	1	$\frac{1}{2}$
C	3	2	1
(d)			
A	1	1	$\frac{1}{2}$
B	1	1	$\frac{1}{2}$
C	2	2	1

Table 4.9 Computation of normalized weights for level 1, illustrated in Figure 4.4

Criteria	*Geometric mean*	*Normalized weights*
Price	$(1 \times 3 \times 1 \times 3)^{1/3} = 1.4422$	0.43
Delivery	$(\frac{1}{3} \times 1 \times \frac{1}{3} \times 1)^{1/3} = 0.4808$	0.14
Quality	$(1 \times 3 \times 1 \times \frac{1}{2})^{1/3} = 1.4422$	0.43
Total	3.3652	1.00

weights for the components, dimensions and vendors is shown in Figure 4.4.

The final step involves combining the weights of the lower level to the next higher level and so on up to the highest level. For instance, the overall ranking for vendor A is computed as follows. The price rating for vendor A is 0.157. The delivery rating for vendor A is obtained by multiplying the lead time rating of vendor A by the weight for lead time and adding it to the product of the lead time variability rating of vendor A and the weight of lead time variability $[(0.659 \times 0.83) +$

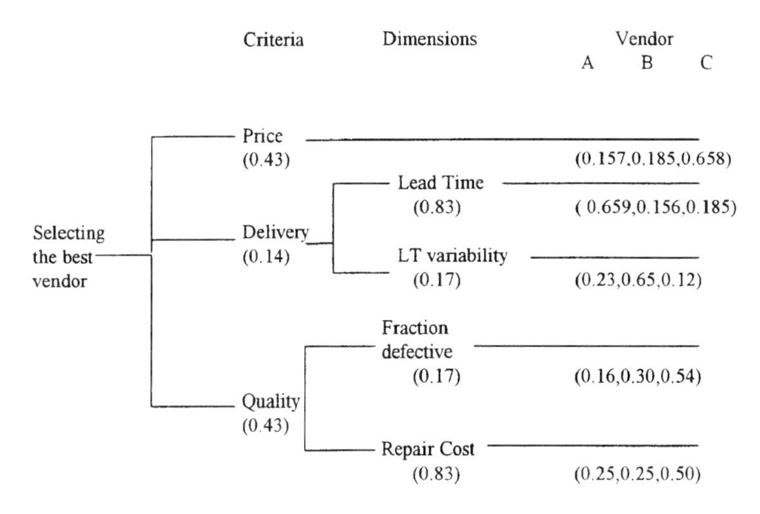

Figure 4.4 Weights (in parentheses) for the hierarchical model.

$(0.23 \times 0.17) = 0.5861]$. The quality rating for vendor A is similarly computed as 0.2347. The overall rating for vendor A is computed by adding the ratings of vendor A for price, delivery and quality with their appropriate weights $[(0.43 \times 0.157) + (0.14 \times 0.5861) + (0.43 \times 0.2347) = 0.25]$.

Similar computations yield the overall ratings for vendors B and C as 0.3141 and 0.4499, respectively. Based on the overall rating, it is evident that vendor C is the best in terms of price, delivery and quality.

4.4 OPTIMIZATION METHODS

Optimization methods use linear or non-linear programming methods to select vendors and determine the order quantities from each vendor. The advantages of using optimization methods are the ability to model different objectives, consider multiple items and address simultaneously certain constraints. For instance, the empirical methods assume that vendors have unlimited capacities and the number of good parts in the supplied lot will meet the demand. The optimization methods can also be used to seek a trade-off between different costs. Current and Weber (1994) demonstrated the applicability of three different facility location models to vendor selection problems. In this section the applicability of two of the facility location models is presented: the simple plant location model (Balinski, 1965) and the set covering location model (Toregas and Re Velle, 1972). In addition, a simple model to consider quality costs is discussed.

4.4.1 Simple plant location model

The simple plant location model can be used to select vendors, as discussed below. The model may be formulated as follows:

$$\text{minimize } z = \sum_k \sum_j c_{kj} X_{kj} + \sum_j f_j Y_j \tag{4.3}$$

subject to

$\sum_j X_{kj} = 1, \forall k;$
$X_{kj} \leq Y_j, \forall k, j;$
$Y_j \in (0, 1), \forall j;$
$X_{ik} \geq 0, \forall k, j.$

where

X_{kj} is the fraction of item k's demand purchased from vendor j;

$$Y_j = \begin{cases} 1 & \text{if vendor } j \text{ is selected,} \\ 0 & \text{otherwise;} \end{cases}$$

c_{kj} is the cost of purchasing all of item k from vendor j;
f_j are the fixed costs associated with employing vendor j.

The objective function, [equation (4.3)] minimizes the total cost of purchasing the items from all vendors. The first term on the right-hand side represents the total variable purchasing costs and the second term on the right-hand side represents the fixed cost associated with utilizing the vendors. The fixed costs include order processing cost, contract set-up cost and follow-up costs. The first constraint ensures that the total demand for each item is satisfied. The second constraint ensures that a purchase order is placed with a vendor only if the vendor is selected. The final two constraints represent the binary and non-negativity restrictions of the decision variables.

If the objective is to minimize late deliveries as in the case of just-in-time manufacturing then the following objective function may be used:

$$\text{minimize } z = \sum_k \sum_j a_k p_{kj} X_{kj} + \sum_j f_j Y_j \tag{4.4}$$

where

a_k is the quantity of item k ordered;
p_{kj} is the percentage of orders of item k delivered late by vendor j.

4.4.2 Set covering location model

The applicability of the set covering location model for vendor selection is presented below. It may be formulated as:

$$\text{minimize } z = \sum_j Y_j \tag{4.5}$$

subject to

$\sum Y_j \geq 1, \forall j$
$j \in N_k$
$Y_j \in (0, 1) \forall j.$

Y_j is defined as before in the simple plant location model and N_k is the set of vendors who can supply item k. The objective function [equation (4.5)] minimizes the number of vendors employed. The first constraint ensures that each item will be supplied by at least one vendor and the second constraint represents the binary nature of the decision variables.

The models presented above assume that the vendors have no capacity limitations. Also, they do not address the quality of the supplied items from the vendors. The set covering model also assumes that the costs related to purchasing are the same for all vendors. The extensive literature available in the area of facility location modeling provides other modeling alternatives to overcome these assumptions. The most serious limitation of the above models is the consideration of a single vendor selection objective. In practice, vendor selection problems are typically multi-objective in nature. Buffa and Jackson (1983) demonstrated the usefulness of a goal programming model to address multi-objective purchasing planning decisions. A more detailed discussion on goal programming formulations can be found in Winston (1987).

4.4.3 Quality cost model

An important consideration in selecting vendors is the quality of supply. Poor supply quality results in additional costs at the buyer end in terms of increased inspection, the repair of defective parts and downtime and/or poor final product quality. An economic model for selecting vendors considering quality costs is discussed in this section (Tagaras and Lee, 1996). The model assumes that the output or finished product of the buyer will be defective if the input or the purchased component is defective, or if the buyer's manufacturing process is defective, or both. The probabilities of a defective input and defective process are assumed to be independent. The model considers the costs of defective finished products and the cost of purchased inputs. All costs are assumed to be linear. The model does not consider incoming inspection on the parts received from the vendors.

The notation is as follows:

p_j is the probability of a defective part from vendor j;
q is the probability of the buyer's process manufacturing a defective finished product;
c_{d} is the unit cost of a defective finished product if the defect was due to defective input (the manufacturing process is perfect);
c_{m} is the unit cost of a defective finished product if the defect was due to a defective manufacturing process (non-defective input);
c_{dm} is the unit cost of a defective finished product if the defect was due to a defective manufacturing process and a defective input;
c_j is the unit cost of purchasing from vendor j;
$p_j(1-q)$ is the probability of producing a defective finished product if the defect was due to a defective input from vendor j (perfect manufacturing process);
$q(1-p_j)$ is the probability of producing a defective finished product if the defect was due to a defective manufacturing process (non-defective input from vendor j);
qp_j is the probability of producing a defective finished product if the defect was due to a defective manufacturing process and a defective input from vendor j;
C_j is the expected cost per unit from vendor j and is the unit purchase cost plus the expected cost of defective finished product [equation (4.6)];
C_0 is the expected cost of a defective finished product if the inputs were perfect [equation (4.7)];
ΔC_j is the incremental cost per unit due to defective parts from vendor j [equation (4.8)].

The model is formulated in terms of the following three equations:

$$C_j = c_j + p_j(1-q)c_{\mathrm{d}} + q(1-p_j)c_{\mathrm{m}} + qp_jc_{\mathrm{dm}} \tag{4.6}$$

$$C_0 = qc_{\mathrm{m}} \tag{4.7}$$

$$\begin{aligned} \Delta C_j &= C_j - C_0 \\ &= c_j + p_j[c_{\mathrm{d}} + q(c_{\mathrm{dm}} - c_{\mathrm{d}} - c_{\mathrm{d}})] \end{aligned} \tag{4.8}$$

Equation (4.8) is evaluated for each vendor and the vendor with the lowest total cost is selected. Appropriate optimization techniques must be used, depending upon the functional form of the costs. It is apparent from equation (4.8) that the total cost increases with decrease in quality of supply from a vendor. Also, if the quality costs are independent of the source of the defect or imperfection ($c_{\mathrm{d}} = c_{\mathrm{m}} = c_{\mathrm{dm}} = c$), then $\Delta C_j = c_j + p_j(1-q)c$. This indicates that if the manufacturing process is perfect then the total cost from vendor j is $c_j + p_jc$. If the manufacturing process is totally imperfect, then the total cost from vendor j is simply the purchase cost, and the quality cost is zero. This implies that when the manufacturing process is not perfect, quality problems associated with supply are less significant.

The model can also be extended to include other costs such as transportation and inventory carrying costs. Additional constraints can also be added to represent capacity restrictions of the vendors. Other enhancements may also include considering the costs of inspecting the supply from the vendors at the buyer site.

4.5 SUMMARY

Vendor selection is one of the most important logistics decisions. The purchase price of items from vendors has a direct impact on the net revenue. The quality of supplied items and delivery reliability affect the uptime of the buyer's production line as well as the quality of the end products. The location of vendors has a great influence on the supply chain design in terms of transportation and distribution planning. In this chapter, several vendor selection models have been discussed. The AHP model discussed here is useful in estimating the relative weights or importance of the various criteria used in the evaluation process. Most companies use some form of weighting scheme to select vendors. Optimization models have not yet found their place in industry, although a very few applications of linear or goal programming methods have been reported in the literature; this is mainly because no generic model can be used for the selection of vendors. The type of model depends to a great extent on the specific application. It should be acknowledged that there is good potential for developing generic optimization based tools specific to an industry.

PROBLEMS

4.1 What is the role of a purchasing department in a company? How does the purchasing function fit into the broad function of logistics?

4.2 Discuss some of the costs associated with vendor selection.

4.3 Discuss the different categories of vendor selection decisions.

4.4 Mention some of the other logistics decisions for which factor analysis and weighted factor methods can be used.

4.5 Formulate a generic goal programming model for vendor selection considering price, quality, delivery, capacity and demand requirements.

4.6 Formulate a transportation model for selecting vendors considering purchase price, transportation cost, demand for items and capacities of vendors.

4.7 An automobile manufacturer is interested in analysing the performance of three major suppliers of their airbags. The purchasing department believes that price, lead time, quality and just-in-time delivery are the four most important factors in selecting a vendor. The normalized range for the values of all the factors is normally distributed with mean 10 and variance 2. Use Monte Carlo simulation (200 iterations) to develop vendor profiles.

4.8 A retail company is in the process of selecting a vendor for supplying ceiling fans. The materials management group has identified price, quality, warranty, packaging, order quantity and weight as key factors. Use the AHP method and your perception of the importance of the factors to determine the weights for these factors.

4.9 In problem 4.8, suppose the ratings for three vendors on the six factors are as follows: Vendor 1, 7,8,6,8,9 and 10; Vendor 2, 8,7,9,5 and 9; and Vendor 3, 9,8,6,6 and 8. Select the best vendor using the weights computed in problem 4.8.

REFERENCES

Balinski, M.L. (1965) Integer programming: methods, uses, computation. *Management Science*, **12**, 253–313.

Ballou, R.H. (1992) *Business Logistics Management*, Prentice-Hall, Englewood Cliffs, NJ.

Buffa, F.P. and Jackson, W.M. (1983) A goal programming model for purchase planning. *Journal of Purchasing and Materials Management*, 27–34.

Burton, T.T. (1988) JIT/repetitive sourcing strategies: tying the knot with your suppliers. *Production and Inventory Management Journal*, 38–41.

Current, J.R. and Weber, C.A. (1994) Application of facility location modeling constructs to vendor selection problems. *European Journal of Operational Research*, **76**, 387–92.

Dickson, G.W. (1966) An analysis of vendor selection systems and decisions. *Journal of Purchasing*, **2**, 5–17.

Lee, L., Jr, Burt, D.N. and Dobler, D.W. (1990) *Purchasing and Materials Management*, McGraw-Hill, New York.

Leenders, M.R. and Fearon, H.E. (1993) *Purchasing and Materials Management*, Irwin, Burr Ridge, IL.

Narasimhan, R. (1983) An analytical approach to supplier selection. *Journal of Purchasing and Materials Management*, 27–32.

Soukup, W.R. (1987) Supplier selection strategies. *Journal of Purchasing and Materials Management*, 7–12.

Tagaras, G. and Lee, H.L. (1996) Economic models for vendor evaluation with quality cost analysis. *Management Science*, 1531–43.

Thompson, K.N. (1990) Vendor profile analysis. *Journal of Purchasing and Materials Management*, 11–18.

Toregas, C. and Re Velle, C. (1972) Optimal location under time or distance constraints. *Papers of the Regional Science Association*, **28**, 133–43.

US Bureau of the Census (1983) *Annual Survey of Manufacturers*, US Government Printing Office, Washington, DC.

Wind, Y. and Robinson, P. (1968) The determinants of vendor selection: the evaluation function approach. *Journal of Purchasing*, **4**, 29–46.
Winston, W.L. (1987) *Operations Research: Applications and Algorithms*, PWS-KENT Publishing, Boston, MA.

FURTHER READING

Gregory, R.E. (1986) Source selection: a matrix approach. *Journal of Purchasing and Materials Management*, 24–9.
Hinkle, C.L., Robinson, P.J. and Green, P.E. (1969) Vendor evaluation using cluster analysis, *Journal of Purchasing*, 49–58.
Ho, C. and Carter, P.L. (1988) Using vendor capacity planning in supplier evaluation. *Journal of Purchasing and Materials Management*, 23–9.
Kasilingam, R.G. and Lee, C.P. (1996) Selection of vendors: a mixed integer programming approach. *Proceedings of the 19th International Conference on Computers and Industrial Engineering*, 234–45; copy available from author.
Narasimhan, R. and Stoynoff, L.K. (1986) Optimizing aggregate procurement allocation decisions. *Journal of Purchasing and Materials Management*, 23–30.
Pan, A.C. (1989) Allocation of order quantities among suppliers. *Journal of Purchasing and Materials Management*, 36–9.
Tang, K. (1988) An economic model for vendor selection. *Journal of Quality Technology*, 81-89.

CHAPTER 5

Inventory planning

Inventory planning and control is one of the logical decisions that follows vendor selection. Once vendors and the total amount of order are determined the next step is to determine the economical order quantity and the ordering frequency. The economic order quantity (EOQ) should be determined based on total costs and service levels. The objectives of minimizing total inventory costs and maximizing service levels are conflicting in nature. Maximization of service level relates to providing a variety of items and ensuring their availability at all times and at all places. This requires maintaining a large inventory of multiple items at several locations close to consumers. Hence, an economic trade-off is needed between inventory costs and service levels.

Inventory is a necessary evil for the proper functioning of the supply chain. Different organizations carry different types of inventories depending upon the nature of their business. Retail companies predominantly maintain finished product inventories. Manufacturing companies may have raw material, work-in-process and finished goods inventories. Service organizations such as hospitals may maintain a large volume of supplies. A work-in-process inventory in manufacturing situations is very similar to a pipeline inventory in distribution or retail situations. Depending on the stage in the supply chain inventories may be classified into raw material, work-in-process, finished goods and scrap (waste). This chapter presents the important functions of inventory and the various costs associated with inventory. A taxonomy of inventory control decisions is then presented. Finally, analytical models to make key inventory control decisions are presented. The models include traditional and just-in-time planning models.

5.1 FUNCTIONS OF INVENTORY

Maintaining the appropriate levels of inventory helps an organization to achieve several important functions. One of the important functions of inventory is to meet uncertainties in production and supply in terms of time, place, variety and quantity. If everything works as planned or forecasted – if the suppliers deliver at the right time at the right place and the supply quality is 100%, if demand forecasts are perfect in terms of time and place for all the items, and if transportation alternatives provide reliable service all the time – then inventory is not needed at all. Unfortunately, trains derail, blizzards hit the highways, customers change their mind at the last minute and shipments get damaged in transit.

Inventory helps to achieve economies of scale in production. (Average or unit production or purchase costs decrease with larger order size owing to economies of scale resulting from fixed costs of ordering and set-up.) Also, several vendors provide quantity discounts for larger order sizes. The use of production with a larger lot size may also lead to better quality or fewer rejects as the learning process is uninterrupted. Larger order sizes may also enable the vendor or buyer to negotiate reduced transportation rates resulting in lower transportation costs. Sufficient care should be exercised to make sure that inventory costs do not offset the benefits of economies of scale.

To protect against anticipated increases in purchase prices or reduction in supply, sometimes it may be worthwhile to incur additional inventory carrying costs, particularly when the purchase prices are expected to increase rapidly and steeply. Also, anticipated strikes at supplier facilities or by transportation carriers may also force buyers to stock up for the future.

Seasonal fluctuations in demand combined with limitations in supply or production capacity also lead to the necessity of maintaining inventories. It may be required to stock up during slow periods in order to meet peak period demand. Sufficient capacity may not be available to purchase or produce peak period requirements. Also, demand for certain items may be time-sensitive. Stocking up of these items is necessary to ensure maximum service levels.

When two adjacent entities in the supply chain operate at different capacities or if the processing or production rates are different for successive production stages, inventory may be the solution to provide independence or balance between the entities or stages. In a distribution situation, the consumption rate and the production rate may be different; an intermediate warehouse may be the answer. In a manufacturing situation, buffer storage between work stations may address imbalance in production rates.

5.2 COSTS OF INVENTORY

There are several different costs associated with maintaining inventory levels: ordering cost, purchase price, storage cost, carrying cost, shortage cost, transportation cost, in-transit inventory cost and quality costs. Although some of the costs increase with higher inventory levels others decrease; hence, a trade-off is often needed to determine the optimum inventory level. Some of the costs of inventory depend upon the lead time to purchase items. Lead time is the time between placing the order and the actual receipt of items. It consists of order preparation and transmittal time, time to manufacture or assemble the item, transportation time and receipt and inspection time. Lead time determines the minimum inventory to be maintained to prevent shortages. Lead time also influences pipeline or in-transit inventory costs.

5.2.1 Procurement and production cost

The procurement cost consists of two components. The first component is the fixed ordering cost, which includes the costs associated with placing the order such as costs of preparing, transmitting and following up the order. This component is typically independent of the order size. The second component is the variable cost of procurement, which is basically the purchase price. This component depends upon the order size and may vary linearly or non-linearly with order size. Non-linearity in purchase price exists when quantity discounts are available either in the form of incremental or all-units discounts. Under all-units discounts, the discount price is applicable to all the units purchased when order size exceeds a certain amount. Under incremental units discount the discounted price is applicable only to the units in that quantity range. The total procurement cost per order is given by:

$$\text{total procurement cost per order} = \text{cost per order} + \text{unit price} \times \text{order size}$$

An illustration is provided in Example 5.1.

Production cost is similar to procurement cost but applies to manufacturing situations. The fixed component is known as the set-up cost, which includes the cost of major and minor set-ups. Typically, set-up costs consists of costs of setting up the job, mounting the tools and of scheduling. The variable component is the variable labor, material and overheads associated with the actual manufacturing.

5.2.2 Holding cost

This is the cost of maintaining inventory. It has two components: storage cost and inventory carrying cost. The first component includes expenses related to storage space, insurance, taxes, obsolescence, heating and lighting. The inventory carrying cost consists of the interest charges related to the investment tied up in inventory and depends upon the interest rate, unit price of the item and average inventory. In most cases storage cost is very low compared with carrying cost. The total holding cost and the cost components are given by:

$$\text{holding cost} = \text{storage cost} + \text{carrying cost}$$

$$\text{storage cost} = (\text{cost per unit}) \times (\text{average inventory})$$

$$\text{carrying cost} = (\text{interest rate}) \times (\text{unit price}) \times (\text{average inventory})$$

When storage costs are negligible compared with carrying costs, the holding cost is equal to the holding cost per unit multiplied by the average inventory, where the holding cost per unit is simply the interest rate multiplied by the unit price. The average inventory is computed based on the beginning and ending inventory when demand is uniform (Figure 5.1).

$$\text{average inventory} = \tfrac{1}{2}(\text{beginning inventory} + \text{ending inventory})$$

EXAMPLE 5.1

CaterMe Corporation requires 2000 units of taco shells every week. CaterMe wants to place a single order every week for 2000 taco shells. The price schedule for taco shells from two vendors is the same and is given in the table below.

Number of shells	*Price per 100 shells* ($)
0–500	25.00
501–1000	22.00
1001–2000	20.00
>2000	18.00

Vendor A is offering all-units discount and vendor B is offering incremental units discount. Compute the procurement costs for placing the order with the two vendors.

Vendor A:

$$\text{2000 shells at a price of \$20 per 100 shells} = 2000 \times \frac{\$20}{100} = \$400$$

Vendor B:

$$\text{first 500 shells at \$25 per 100 shells} = 500 \times \frac{\$25}{100} = \$125$$

$$\text{next 500 shells at \$22 per 100 shells} = 500 \times \frac{\$22}{100} = \$110$$

$$\text{last 1000 at \$20 per 100 shells} = 1000 \times \frac{\$20}{100} = \$200$$

$$\text{total cost} = \$125 + \$110 + \$200 = \$435$$

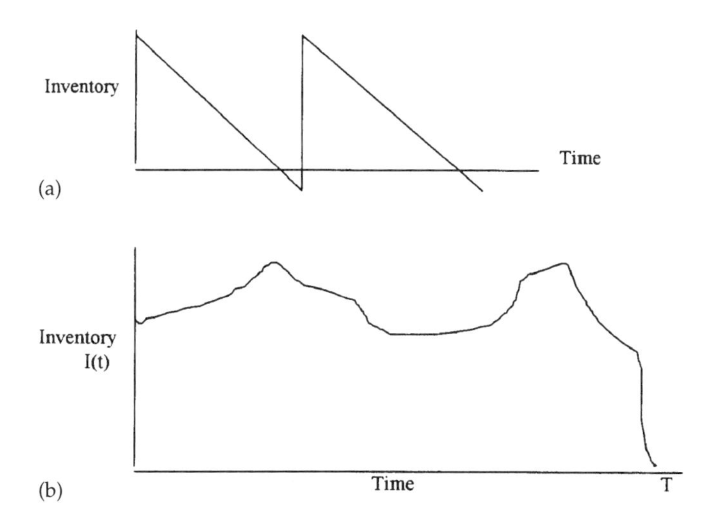

Figure 5.1 Demand: (a) uniform; (b) continuous, non-uniform.

When demand is continuous [Figure 5.1(b)] the average inventory can be computed by means of integration, as follows:

$$\text{average inventory} = \frac{1}{T}\int_0^T I(t)\mathrm{d}t$$

5.2.3 Shortage cost

When demand is lost or production is stopped owing to lack of product or material availability certain costs are incurred. In a sales or distribution environment, this leads to lost sales resulting in loss of revenue or profit and goodwill. When items can be backordered this results in additional expenses to fill the demand expeditiously. This may include buying the items from other vendors, using premium transportation and order follow-up. In a manufacturing situation, this results in expenses related to machine downtime and shortage of end products for sale. When sales are lost shortage cost is a function of units of demand not satisfied. When units can be backordered shortage cost is a function of the number of units backordered and the time to fill the order:

$$\text{shortage (lost sales)} = (\text{profit and goodwill cost per unit}) \times (\text{units lost})$$

$$\begin{aligned}\text{shortage cost (backordering)} = {} & (\text{backorder cost per unit per unit time}) \\ & \times (\text{units backordered}) \\ & \times (\text{length of backorder})\end{aligned}$$

5.2.4 Transportation and in-transit inventory costs

In general, the transportation cost depends upon the type of commodity, distance, mode of transportation, carrier and shipment size. The in-transit inventory carrying cost depends upon the amount of inventory in the transportation pipeline and the number of days in the pipeline.

As speed of transportation increases (as a result of the mode selected) the cost of in-transit inventory goes down but the cost of transportation goes up. For example, whereas air transportation may take only a day to move shipments from Los Angeles to New York, rail transportation may take more than seven days. Obviously, the cost of shipping by airplanes is much higher compared with the cost of moving by rail, but the inventory carrying cost is less. Larger order or shipment sizes result in lower transportation costs – for example, as when truckload carriers are used instead of less-than-truckload carriers.

5.2.5 Quality costs

These are the costs associated with receiving defective parts in a shipment. A part in a shipment may be defective as a result of poor manufacturing, wrong specifications or damage during transit and handling. Consequently, the defective parts contribute to problems at the buyer site. Non-availability of the required number of good parts causes plant downtime and results in not meeting the final product demand. Defective parts passing through inspection at buyer site leads to a poor quality final product. When defectives are found during inspection additional costs have to be incurred to repair the parts or to ship them back to the suppliers. In general, inventory order size should be adjusted to account for defective parts. Quantification of quality costs is often difficult. Also, it depends upon the type of inspection and sampling plans used by the buyer.

5.3 CLASSIFICATION OF INVENTORY CONTROL PROBLEMS

This section presents a classification of the different types of inventory control decisions. Inventory decisions may be classified in several ways. In this section, inventory control problems are classified based on the number of locations, number of supply sources, number of items, number of periods, nature of demand and type of lead time.

Certain items may be stocked in several locations. The locations may be in different buildings within the same plant or in different cities within a region or may be in different countries. The location of the inventory has a great influence on transportation costs and selection of vendors. The combining of demand from nearby locations can result in procurement and transportation economies of scale. Different locations may have to adopt different types of inventory policies since demand and costs differ. The number of supply sources affects the degree of supply reliability. In some cases it may be desirable to distribute demand to several vendors to ensure availability on time. On the other hand, this may cause variability in the quality of incoming parts and lead to substitutability problems. Increase in the number of items leads to greater efforts in terms of operating, maintaining and controlling inventories. However, the benefits may include greater opportunities for quantity discounts.

Certain types of inventory control decisions are related to single periods or non-repetitive orders such as special seasonal or holiday demand and other one-time demands. Examples of this include Christmas items stocking, inventory for construction materials to build a block of houses and special services for a function. Multiple period inventory planning may include repetitive and non-repetitive ordering. If the demand during all the periods is the same then orders are placed at fixed intervals for the same order size. Otherwise, order size and/or ordering frequency may vary.

The nature of demand greatly influences the type of inventory system and the control policies. The nature of demand has three dimensions: dependency, variability and uncertainty. When multiple items are present the dependency of demand among the items is a key factor. If the demands for the various items are independent then the demand for each item must be individually forecasted. If dependency exists between items then the demand for one item may be derived from the demand for another item. A good example of this is the relationship between an end product and its components or subassemblies. For instance, once the demand for automobiles is known, it is easy to derive the demand for axles, engines, etc. Demand variability refers to the variation of demand over time. Constant demand inventory problems are easy to handle compared with variable demand problems. Similarly, deterministic demand situations are easy to model and analyse compared with stochastic inventory problems. The complexity involved in analysing stochastic demand inventory problems depends on the type of underlying distribution.

5.4 BASIC INVENTORY CONTROL MODELS

This section presents two models. The first one addresses a single period situation using the classical economical order quantity (EOQ) approach. The second addresses a multiperiod situation using a linear programming approach. The models presented in this section assume the following:

1. demand and lead time are known with certainty;
2. all costs are linear and known;
3. no stockouts are allowed.

The following costs are considered: ordering cost, purchase price, inventory holding cost, transportation cost and in-transit inventory cost.

5.4.1 Single-period model

The notation is as follows:

A is the ordering cost;
C is the unit purchase price;
D is the annual demand;
L is the lead time, in days;
Q is the order quantity;
i is the interest rate per year;
R is the reorder point;
s is safety stock;
t is the transit time in days;
T is the unit transportation cost;
S is the storage cost per unit;
h is the holding cost per unit and is given by $i(C+T)$;
C^{tot} is the total annual cost.

The model may be specified as follows:

$$\begin{aligned}\text{total cost} = {} & \text{ordering cost} + \text{purchase cost} + \text{inventory carrying cost} \\ & + \text{transportation cost} + \text{in-transit inventory cost}\end{aligned}$$

$$\begin{aligned}\text{annual ordering cost} &= (\text{cost per order}) \times (\text{number of orders per year}) \\ &= \frac{AD}{Q}\end{aligned}$$

$$\begin{aligned}\text{annual purchase cost} &= (\text{cost per unit}) \times (\text{demand per year}) \\ &= CD\end{aligned}$$

$$\begin{aligned}\text{annual holding cost} &= (\text{average inventory}) \times [(\text{storage cost per unit}) \\ &\quad + (\text{cost per unit} + \text{transportation cost per unit}) \\ &\quad \times \text{interest rate})] \\ &= \left(\frac{Q}{2}\right)[S + (C+T)i] \\ &= \left(\frac{Q}{2}\right)h, \text{ when } S \text{ is insignificant compared} \\ &\quad \text{with } (C+T)i\end{aligned}$$

$$\text{transportation cost} = (\text{transportation cost per unit}) \times (\text{annual demand}) = TD$$

$$\text{in-transit inventory cost} = (\text{interest rate per year}) \times (\text{unit purchase price}) \times (\text{annual demand}) \times (\text{transit time in years}) = \frac{iCDt}{365}$$

$$C^{\text{tot}}(Q) = \frac{AD}{Q} + CD + \left(\frac{Q}{2}\right)[S + (C+T)i] + TD + \frac{iCDt}{365} \tag{5.1}$$

To obtain the EOQ, the value of Q that minimizes the total cost, one must take the first derivative of the total annual cost expression given in equation (5.1) with respect to Q, set it equal to zero, and solve for Q. This results in the following formula for the EOQ:

$$Q^* = \left[\frac{2AD}{i(C+T)+S}\right]^{1/2} \tag{5.2}$$

$$Q^* = \left(\frac{2AD}{h}\right) \text{ when } S \text{ is very small} \tag{5.3}$$

The reorder point for this situation is simply equal to the lead time demand plus safety stock:

$$R = \frac{LD}{365} + s \tag{5.4}$$

The transportation inventory costs do not seem to impact the EOQ. The primary reason is that in this model they do not depend upon the order quantity, and the transportation costs and transit times are not linked to each other. In other words, an increase or decrease in transit time does not have any impact on the transportation cost. The model can be modified to consider quantity discounts on transportation rate and/or purchase price. Additional constraints may also be added to the model to consider maximum storage space, budget availability, etc. The model can also be extended to consider non-linear transportation costs and to include multiple vendors with a common reordering schedule. For an extensive treatment of all types of EOQ inventory models the reader is referred to Tersine (1994). An illustration is given in Example 5.2.

5.4.2 Multiperiod models

Linear programming model

The notation is as follows:

X_t is the quantity ordered and received in period t (the lead time is assumed to be zero);
I_0 is the inventory at the beginning of the planning horizon;
I_t is the inventory at the end of period t;
I_n is the desired ending inventory;
C_t is the purchase price in period t;
T_t is the cost of transportation in period t;
h_t is the holding cost in period t (including interest and storage cost);
D_t is the demand in period t;
S_t is the availability of supply in period t;
V^{st} is the storage capacity.

The mathematical representation of the model is as given below:

$$\text{minimize } Z = \sum_t X_t(C_t + T_t) + \sum_t I_t h_t \tag{5.5}$$

subject to

$$\begin{aligned} (X_t + I_{t-1}) &\leq V^{st}, \forall t \\ I_{t-1} + X_t - D_t &= I_t, \forall t \\ X_t &\leq S_t, \forall t \\ I_t, X_t &\geq 0, \forall t \end{aligned}$$

EXAMPLE 5.2

XYZ Corporation buys one of the key components of its final product from ABC Corporation. The component has an annual demand of 8000 units. The cost of placing an order is \$40.00. The unit purchase price, unit transportation cost and unit storage cost of the item are \$5, \$0.50 and \$0.10, respectively. The transit time is 10 days, and the lead time is about 15 days. Assuming an interest rate of 10%, compute the economic order quantity Q and reorder point R.

Substitute the values into equation (5.2), to give:

$$Q^* = \left(\frac{2 \times 40 \times 8000}{0.1(5 + 0.5) + 0.1}\right)^{1/2} \text{ units}$$

$$Q^* \simeq 992 \text{ units}$$

The reorder point R is given by the lead time multiplied by the number of units needed per day:

$$\begin{aligned} R &= \frac{15 \times 8000}{365} \text{ units} \\ &= 328 \text{ units} \end{aligned}$$

The objective function of the model, [equation (5.5)] minimizes the total cost related to the inventory planning situation. The first term on the right-hand side represents the sum of purchase cost and transportation cost. The second term on the right-hand side represents the total holding cost incurred. The constraints include storage capacity restrictions and inventory balance constraints. The first constraint ensures that the maximum inventory level does not exceed the available storage capacity. The second constraint indicates that the inventory at the end of period t is given by the inventory at the end of period $(t-1)$ plus the quantity received in period t minus the quantity consumed in period t. The assumption here is to meet the demand in every period without any shortages. The third constraint limits the order quantity to the available supply. The final constraint represents the non-negativity of the decision variables.

The above model can be easily extended to consider multiple items, multiple supply sources and backordering situations. For a thorough discussion on mathematical models to address a variety of multiperiod inventory situations the reader is referred to Johnson and Montgomery (1974). Multiperiod models with linear costs may be represented in the form of a transportation model and solved efficiently by means of transportation algorithms.

Transportation model

In this section the transportation modeling of the multiperiod inventory problem is illustrated with use of Example 5.3. For a detailed description of the formulation the reader is referred to Winston (1987).

The transportation model shown in Example 5.3 can be extended to consider backordering and to include multiple items, multiple vendors and multiple locations. The model given in the transportation table can be solved efficiently by using specialized techniques. The first step is to obtain a good feasible solution. Several procedures are available for this, including the Northwest corner rule and Vogel's approximation method. The second step is to continue from the feasible solution to determine the optimal solution. Vogel's method is illustrated in this section. A description of the transportation simplex procedure to determine the optimal solution can be found in Winston (1987).

Vogel's method

Vogel's method may be described as follows.

Step 1: add a dummy row or column to balance the transportation table. A dummy row or column will have zero cost coefficients in the cells. If total demand is greater than total supply, add a dummy row with supply equal to total demand less total

EXAMPLE 5.3

Swings Enterprises sells a particular class of children's swing sets. The most popular model, ZX3, has demand throughout the year with some seasonality. The inventory holding cost is \$3 per unit per quarter. Assume that demand cannot be backordered and shortages are not allowed. The purchase cost and transportation rates vary depending upon the time of the year. The quarterly demand for ZX3, purchase price and the transportation costs for the next four quarters are given in the following table.

	Quarter			
	1	*2*	*3*	*4*
Demand	600	500	600	700
Capacity	800	500	400	700
Purchase price (\$/unit)	180	181	182	183
Transportation cost (\$/unit)	4	5	5	6

This is a balanced transportation problem since the total demand and the total supply are equal. The transportation table for the problem is as follows.

Quarter	*Quarter*				*Supply*
	1	*2*	*3*	*4*	
1	\$184	\$187	\$190	\$193	800
2	\$$\infty$	\$186	\$189	\$192	500
3	\$$\infty$	\$$\infty$	\$187	\$190	400
4	\$$\infty$	\$$\infty$	\$$\infty$	\$189	700
Demand	600	500	600	700	

The last column indicates the availability of supply in each quarter. The last row indicates the demand in each quarter. The entries in other cells indicate the costs corresponding to certain inventory decisions. The entry in cell (1, 1) which corresponds to (quarter 1, quarter 1) shows that the cost of buying in quarter 1 to meet the demand in quarter 1 is equal to the purchase cost plus the transportation cost in quarter 1 (\$180 + \$4). The entry in cell (2, 4) which corresponds to (quarter 2, quarter 4) shows that the cost of buying in quarter 2 to meet the demand in quarter 4 is equal to the purchasing and transportation cost in quarter 2 and the cost of carrying inventory for two quarters, from quarter 2 to quarter 4 (\$181 + \$5 + 2 × \$3 = \$192). The extremely high cost, ∞, in cell (3, 1) indicates that demand for quarter 1 cannot be met by buying in quarter 3.

supply. If total supply is greater than total demand, add a column with demand equal to total supply minus total demand.

Step 2: for each row compute the difference between the two smallest costs. Similarly, compute the difference between the two smallest costs for each column. For Example 5.3 this will result in row differences of 3 for all rows, and column differences of ∞, 1, 1, and 1.

Step 3: choose the row or column corresponding to the largest difference and allocate the minimum of supply and demand to the cell with the least cost. When the largest difference occurs in more than one row or column break the tie by selecting the column or row that contains a cell with the least cost. When the least cost occurs for several cells, break tie arbitrarily. In our example, column 1 and row 4 have the largest difference; however, the cell with the least cost occurs in column 1, cell (1,1). The supply and demand corresponding to cell (1,1) are 800 and 600, respectively. Allocate 600 (min {800,600}) units to cell (1,1).

Step 4: modify the matrix based on the allocation. If demand is met, delete the column and compute the remaining supply. If supply is used up, delete the row and adjust the demand. In our example, column 1 is deleted since its demand is satisfied and supply available in row 1 is now reduced from 800 to 200 (800 – 600).

	Quarter				
Quarter	*1*	*2*	*3*	*4*	*Supply*
1	184^{(600)}$	$187	$190	$193	200
2	$$\infty$	$186	$189	$192	500
3	$$\infty$	$$\infty$	$187	$190	400
4	$$\infty$	$$\infty$	$$\infty$	$189	700
Demand	0	500	600	700	

Step 5: if all the allocations are done, then stop. Otherwise, repeat steps 2 through 4. All allocations are not completed. Repeating steps 2 through 4 on the table given in step 4 results in the following.

	Quarter				
Quarter	*1*	*2*	*3*	*4*	*Supply*
1	184^{(600)}$	$187	190^{(200)}$	$193	0
2	$$\infty$	186^{(500)}$	$189	$192	0
3	$$\infty$	$$\infty$	187^{(400)}$	$190	0
4	$$\infty$	$$\infty$	$$\infty$	189^{(700)}$	0
Demand	0	0	0	0	

The procurement and inventory decision for the above problem is as follows. Buy 800 units in quarter 1 to cover the entire demand for quarter 1 and 200 units of demand for quarter 3. Buy 400 units in quarter 3 to cover the remaining quarter 3 demand. Buy 200 units in quarter 2 to meet quarter 2 demand and 700 units in quarter 4 to cover quarter 4 demand.

5.4.3 General analytical model

In certain inventory control situations it may not be possible to formulate a mathematical model and obtain a closed-form solution. The cost functions may not be linear. The feasible values for order quantity may be finite. However, it is possible to arrive at the best decision analytically. With larger order size, transportation costs and ordering costs decrease; however, inventory costs increase. This is illustrated in Example 5.4.

EXAMPLE 5.4

ABC Corporation buys the engines for its lawn mowers from Engines Unlimited. The annual demand for engines is 4000. The weight of each engine is 200 lb. The purchase price of an engine is $600 and the cost of placing an order is $20. The handling and inspection cost at ABC's warehouse is $0.25 per cwt. Three trucking carriers are available to transport the motors from the vendor site. Carrier 1 can haul a maximum of 10 000 lb at a rate of $9 per cwt. Carrier 2 will haul a maximum of 30 000 lb at a rate of $8 per cwt subject to a minimum of 20 000 lb. Carrier 3 will haul at a rate of $7 per cwt subject to a minimum of 40 000 lb. Assuming unlimited warehouse space and an inventory carrying rate of 25% decide on the best transportation alternative.

The available transportation options and the associated key information are as follows.

Option	Description	Order size (no. engines)	Number of orders	Transport cost ($/cwt)	Average inventory
1	Carrier 1 (10 000 lb)	50	80	9	25
2	Carrier 2 (20 000 lb)	100	40	8	50
3	Carrier 2 (30 000 lb)	150	26.7	8	75
4	Carrier 3 (40 000 lb)	200	20	7	100

For carrier 3 we also have options to transport more than 40 000 lb. We are limiting our options here for simplicity. Handling costs are not considered since they depend upon the total number of engines handled per year and not on the order size.

The sample calculations for option 2 are as follows:

$$\text{ordering cost (\$20 per order, 40 orders per year)} \\ = 40 \times \$20 = \$800$$

$$\text{transportation cost (\$8 per cwt, 200 cwt per order, 40 orders per year)} \\ = 40 \times 200 \times \$8 = \$64\,000$$

$$\text{inventory cost (50 engines at \$600, \$0.25 per dollar per year)} \\ = 0.25 \times 50 \times \$600 = \$7500$$

It should be noted that transportation cost can also be computed directly from the knowledge that one has 4000 engines, each weighing 2 cwt at a charge of \$8 per hundredweight ($4000 \times 2\,\text{cwt} \times \$8/\text{cwt} = \$64\,000$). The resulting cost computations for the four options are given in the table below. Based on the total cost of the four options, option 4 seems to be the best.

Option	Description	Ordering cost (\$)	Inventory cost (\$)	Transport cost (\$)	Total cost (\$)
1	Carrier 1 (10 000 lb)	1600	3750	72 000	77 350
2	Carrier 2 (20 000 lb)	800	7500	64 000	72 300
3	Carrier 2 (30 000 lb)	534	11 250	64 000	75 784
4	Carrier 3 (40 000 lb)	400	15 000	56 000	71 400

5.5 JUST-IN-TIME MODELS

In this section three models representing just-in-time ordering situations are presented. The first model considers blanket ordering in which an order is placed for a certain amount but is delivered by the vendor in small amounts at regular intervals. The model essentially determines the frequency of delivery given the blanket order quantity. The second model determines the order quantity and the delivery frequency. The third model determines the order quantity and delivery frequency considering multiple vendors. In other words, it combines vendor selection and just-in-time inventory control.

5.5.1 Blanket ordering model

In this model the entire demand for a prespecified period (e.g. one quarter) is placed as a blanket order with the actual order delivered in several shipments on an installment basis. The cost of transportation increases as the number of deliveries increases; however, the cost of carrying inventory decreases. The objective is to seek a trade-off between these two costs in order to determine the optimum number of deliveries per year. The notation is as follows:

A is the ordering cost;
C is the unit purchase price;
D is the annual demand;
L is the lead time, in days;
Q is the order quantity;
n is the number of deliveries per order;
N is the number of deliveries per year;
i is the interest rate per year;
R is the reorder point;
s is safety stock;
V is the transportation cost per delivery;
t is the transit time in days;
T is the unit transportation cost;
S is the storage cost per unit;
h is the holding cost per unit, given by $i(C+T)$;
C^{tot} is the total annual cost.
The model may be specified as follows:

$$\begin{aligned}\text{total cost} &= \text{ordering cost} + \text{purchase cost} + \text{inventory carrying cost}\\ &\quad + \text{transportation cost} + \text{in-transit inventory cost}\end{aligned}$$

$$\text{number of deliveries per year} = n\frac{D}{Q}$$

$$\text{size of delivery} = \frac{D}{N}$$

$$\text{average inventory} = \frac{D}{2N}$$

$$\begin{aligned}\text{annual ordering cost} &= (\text{cost per order}) \times (\text{number of orders per year})\\ &= \frac{AD}{Q}\end{aligned}$$

$$\begin{aligned}\text{annual purchase cost} &= (\text{cost per unit}) \times (\text{demand per year})\\ &= CD\end{aligned}$$

$$\text{annual holding cost} = (\text{average inventory}) \times (\text{holding cost per unit}) = \frac{Dh}{2N}$$

$$\begin{aligned}\text{transportation cost} &= (\text{transportation cost per unit}) \times (\text{annual demand}) \\ &\quad + (\text{transportation cost per delivery}) \\ &\quad \times (\text{number of deliveries}) \\ &= TD + VN\end{aligned}$$

$$\begin{aligned}\text{in-transit inventory cost} &= (\text{interest rate per year}) \times (\text{unit purchase price}) \\ &\quad \times (\text{annual demand}) \times (\text{transit time in years}) \\ &= \frac{iCDt}{365}\end{aligned}$$

$$C^{\text{tot}}(N) = \frac{AD}{Q} + CD + \frac{Dh}{2N} + TD + VN + \frac{iCDt}{365} \tag{5.6}$$

To obtain the optimum number of deliveries, the value of N that minimizes the total cost, we take the first derivative of the total annual cost expression given in equation (5.6) with respect to N, set it equal to zero, and solve for N. This results in the following formula for the EOQ:

$$N^* = \left(\frac{Dh}{2V}\right)^{1/2} \tag{5.7}$$

An illustration is provided in Example 5.5.

EXAMPLE 5.5

JIT Corporation buys one of the key components of its final product from GET Corporation. The component has an annual demand of 9000 units. The cost of placing an order is $40.00. The unit purchase price, unit transportation cost and transportation cost per delivery are $5, $0.50 and $10, respectively. Order quantity has been specified by the management as 1000 units. Determine the optimal number of deliveries per order assuming a holding cost of $2 per unit.

$$N^* = \left(\frac{9000 \times 2}{2 \times 10}\right)^{1/2} = 30 \text{ deliveries per year}$$

$$n^* = \frac{QN^*}{D} = \frac{1000 \times 30}{9000} = 3.3 \text{ deliveries per order}$$

5.5.2 Just-in-time economic order quantity model

This model demonstrates the usefulness of the application of the EOQ model to just-in-time situations. The model determines the optimal order quantity and the optimal number of deliveries per order based on the costs of ordering, purchasing, holding and transportation. The cost of transportation increases as the number of deliveries increases; however, the cost of carrying inventory decreases. The objective is to seek a trade-off between these two costs to determine the optimum number of deliveries per year.

Following the same notation as in the blanket ordering model, the just-in-time model is described as follows:

$$\begin{aligned}\text{total cost} &= \text{ordering cost} + \text{purchase cost} + \text{inventory carrying cost}\\ &\quad + \text{transportation cost} + \text{in-transit inventory cost}\end{aligned}$$

$$\text{number of deliveries per year} = \frac{Dn}{Q}$$

$$\text{size of delivery} = \frac{Q}{n}$$

$$\text{average inventory} = \frac{Q}{2n}$$

$$\begin{aligned}\text{annual ordering cost} &= (\text{cost per order}) \times (\text{number of orders per year})\\ &= \frac{AD}{Q}\end{aligned}$$

$$\begin{aligned}\text{annual purchase cost} &= (\text{cost per unit}) \times (\text{demand per year})\\ &= CD\end{aligned}$$

$$\begin{aligned}\text{annual holding cost} &= (\text{average inventory}) \times (\text{holding cost per unit})\\ &= \frac{Qh}{2n}\end{aligned}$$

$$\begin{aligned}\text{transportation cost} &= (\text{transportation cost per unit}) \times (\text{annual demand})\\ &\quad + (\text{transportation cost per delivery})\\ &\quad \times (\text{number of deliveries})\\ &= TD + \frac{VDn}{Q}\end{aligned}$$

$$\text{in-transit inventory cost} = (\text{interest rate per year}) \times (\text{unit purchase price}) \times (\text{annual demand}) \times (\text{transit time in years}) = \frac{iCDt}{365}$$

$$C^{\text{tot}}(Q, n) = \frac{AD}{Q} + CD + \frac{Q}{2n}\left[S + \left(C + T + \frac{V}{Q}\right)i\right] + TD + \frac{VDn}{Q} + \frac{iCDt}{365} \tag{5.8}$$

The total annual cost expressed in equation (5.8) is a function of the order quantity and the number of deliveries per order. A solution to the above expression can be obtained by using an iterative procedure. For a fixed value of Q, expression (5.8) becomes a function of n, $C^{\text{tot}}(n)$. Similarly, for a known value of n, it becomes a function of Q, $C^{\text{tot}}(Q)$.

The iterative procedure is as follows:

Step 1: start with an arbitrary value of Q^{a} and solve $C^{\text{tot}}(n, Q^{\text{a}})$ optimally using calculus to obtain n'. This can be done using the following equation:

$$n' = Q^{\text{a}}\left(\frac{h}{2VD}\right)^{1/2} \tag{5.9}$$

Step 2: solve $C^{\text{tot}}(Q, n')$ optimally for Q' using the following equation:

$$Q' = \left(\frac{2n'D(A + Vn)}{h}\right)^{1/2} \tag{5.10}$$

Step 3: now solve $C^{\text{tot}}(n', Q')$ for n''.

Step 4: repeat steps 2 and 3 until convergence is reached. Convergence is reached if the values of Q and n between two successive iterations are close, or if the total cost between two consecutive iterations are very close.

5.5.3 Just-in-time inventory and vendor selection model (Gnanendran, 1995)

This model is an extension of the previous model to consider multiple vendors. The objective here is to determine the size of the orders to be placed to different vendors and the delivery frequency in order to minimize the costs of purchasing, transportation and inventory holding. The model formulation is based on the notation used in the preceding just-in-time models. An additional subscript j is added to all the variables to indicate vendor j. The model assumes that one blanket order is placed per year with each vendor. The relevant costs are purchase cost, inventory cost, transportation cost and in-transit inventory cost. Thus, following the same notation as in the blanket ordering model, the just-in-time model is described below.

$$\text{total cost} = \text{purchase cost} + \text{inventory carrying cost} \\ + \text{transportation cost} + \text{in-transit inventory cost}$$

$$\text{size of delivery} = \frac{Q_j}{N_j}$$

$$\text{average inventory} = \frac{Q_j}{2N_j}$$

$$\text{annual purchase cost} = (\text{cost per unit}) \times (\text{demand per year}) \\ = \sum C_j Q_j$$

$$\text{annual holding cost} = (\text{average inventory}) \times (\text{holding cost per unit}) \\ = \sum \frac{Q_j h}{2N_j}$$

$$\text{transportation cost} = (\text{transportation cost per unit}) \times (\text{annual demand}) \\ + (\text{transportation cost per delivery}) \\ \times (\text{numbers of deliveries}) \\ = \sum T_j Q_j + \sum V_j N_j$$

$$\text{in-transit inventory cost} = (\text{interest rate per year}) \times (\text{unit purchase price}) \\ \times (\text{annual demand}) \times (\text{transit time in years}) \\ = \sum \frac{Q_j h_j t_j}{365}$$

$$\text{minimize } C^{\text{tot}}(Q_j, N_j) = \sum C_j Q_j + \sum \frac{Q_j h_j}{2N_j} + \sum T_j Q_j \\ + \sum V_j N_j + \sum \frac{Q_j h_j t_j}{365} \tag{5.11}$$

subject to

$$\sum Q_j = D$$

$$Q_j, N_j \geq 0$$

To solve the above model, the following transformation can be made. The problem can be written as:

$$\text{minimize } V(Q_j) = \sum Z(Q_j) + \sum C_j Q_j + \sum T_j Q_j + \sum \frac{Q_j h_j t_j}{365} \tag{5.13}$$

subject to

$$\sum Q_j = D$$
$$Q_j \geq 0$$

$Z(Q_j)$ is given by the optimal objective value of the following unconstrained minimization problem:

$$Z(Q_j) = \min_{N_j} \left\{ \frac{Q_j h_j}{2N_j} + V_j N_j \right\} \tag{5.14}$$

The optimal N_j is given by $(Q_j h_j / 2V_j)^{1/2}$ with the corresponding $Z(Q_j) = (2Q_j h_j V_j)^{1/2}$. Hence, the problem reduces to:

$$\text{minimize } \sum (2Q_j h_j V_j)^{1/2} + \sum C_j Q_j + \sum T_j Q_j + \sum \frac{Q_j h_j t_j}{365} \tag{5.15}$$

subject to

$$\sum Q_j = D$$
$$Q_j \geq 0$$

Let q_j (equal to Q_j/D) denote the fractional demand allocated to vendor j and define b_j as $(2Dh_j V_j)^{1/2}$. The problem can now be written as follows:

$$\text{minimize } \sum b_j (q_j)^{1/2} + \sum C_j D q_j + \sum T_j D q_j + \sum \frac{D q_j h_j t_j}{365} \tag{5.16}$$

subject to

$$\sum Q_j = D$$
$$Q_j \geq 0$$

For a single vendor, this problem can be trivially solved with $q_j = 1$, $Q_j = D$, and $N_j = (Dh_j/2V_j)^{1/2}$. In the case of multiple vendors without capacity restrictions, it can be shown that single sourcing is always optimal. To identify the optimal source, simply rank the vendors in ascending order of the value of $b_j + D(C_j + T_j + q_j h_j t_j / 365)$. Once the optimal source is identified, the optimal number of deliveries can be computed as $(Dh_j/2V_j)^{1/2}$. For a situation with vendor capacity restrictions, the model becomes a non-linear program and may have to be solved by means of non-linear programming packages.

5.6 SUMMARY

The key inventory decisions are when and how much to order. Inventory cost is a significant component of the total logistics cost. Costs of inventory include holding and shortage costs of items in storage and in-transit. However, inventory decisions are influenced by other costs as well. For instance, purchase price and freight rates depend on order and shipment size which in turn affect inventory. In some cases, inventory decisions may simply be to determine when to receive the deliveries for a known or blanket order size to maintain just-in-time control. In this chapter, a few models have been presented that consider a total cost approach to inventory modeling. Some basic aspects of multiperiod inventory modeling have also been discussed. A model to combine inventory and vendor selection decisions was also briefly outlined.

PROBLEMS

5.1 Discuss the different types of inventory and the reasons for maintaining them.

5.2 Sketch the relationship between average inventory and the costs of holding and ordering.

5.3 Discuss the importance of safety stock and the implications of shortage.

5.4 Derive the minimum total cost equation for the single period inventory model discussed in section 5.4.1.

5.5 Develop a linear programming formulation of the transportation inventory control model discussed in section 5.4.2.

5.6 ABC Corporation buys one of the key components of its final product from SupplyPerfect Inc. The component has an annual demand of 9000 units. The cost of placing an order is $30.00. The unit purchase price, unit transportation cost and unit storage cost of the item are $4, $0.40 and $0.10, respectively. The transit time is 20 days and the lead time is about 25 days. Assuming an interest rate of 12% compute the economic order quantity and reorder point.

5.7 Develop a transportation table for the transportation model discussed in section 5.4.2 to include backordering. Assume a backordering cost of $2/unit/quarter.

5.8 A bicycle manufacturing company is interested in optimally planning the inventory levels for tires. Inventory holding cost is $3 per tire per period and the capacity of the vendor is 400 units per period. Assume that demand can be backordered at a cost of $5 per tire per period. The purchase price and transportation rates vary depending upon the time of the year. The demand for tires is derived from the production schedule for bicycles. The purchase price, demand and transportation costs for the next four quarters are as follows.

	Quarter					
	1	*2*	*3*	*4*	*5*	*6*
Demand	500	300	500	400	200	500
Purchase price ($/unit)	80	100	110	90	90	110
Transportation cost ($/unit)	5	6	6	5	6	6

Determine the optimal purchasing and inventory plan.

5.9 For the data given in Example 5.5, verify if the order quantity specified by the management is optimal.

REFERENCES

Gnanendran, S.K. (1995) An integrated model for allocating order quantities and setting delivery frequencies. Research report, School of Management, University of Scranton, Scranton, PA.

Johnson, L.A. and Montgomery, D.C. (1974) *Operations Research in Production Planning and Inventory Control*, John Wiley, New York.

Tersine, R.J. (1994) *Principles of Inventory and Materials Management*, Prentice-Hall, Englewood Cliffs, NJ.

Winston, W.L. (1987) *Operations Research: Applications and Algorithms*, PWS-KENT Publishing, Boston, MA.

FURTHER READING

Ansari, A. and Modarress, B. (1986) Just-in-time purchasing: problems and solutions. *Journal of Purchasing and Materials Management*, 11–15.

Buffa, F.P. and Munn, J.R. (1990) Multi-item grouping algorithm yielding near-optimal logistics cost. *Decision Sciences*, 14–34.

Council of Logistics Management (1994) *Bibliography of Logistics Training Aids*, 2803 Butterfield Road, Suite 38, Oak Brook, IL 60521.

Hong, J.-D. and Hayya, J.C. (1992) Just-in-time purchasing: single or multiple sourcing. *International Journal of Production Economics*, 175–81.

Kasilingam, R.G. (1997) An economic model for overbooking of inventory under uncertainty. *Computers and Industrial Engineering: An International Journal*, 221–226.

Kasilingam, R.G. and Tabucanon, M.T. (1987) Integrated production and sales planning in a paper mill using linear programming. *Modern Production Management Systems*, Elsevier, New York, pp. 257–64.

Langley, C.J. Jr (1981) The inclusion of transportation costs in inventory models. *Journal of Business Logistics*, **2**, 106–25.

Pan, A.C. and Liao, C.-J. (1989) An inventory model under just-in-time purchasing agreements. *Production and Inventory Management Journal*, 49–52.

Ramasesh, R.V. (1990) Recasting the traditional inventory model to implement just-in-time purchasing. *Production and Inventory Management Journal*, 71–5.

Russell, R.M. and Krajewski, L.J. (1991) Optimal purchasing and transportation cost lot sizing for a single item. *Decision Sciences*, 940–52.

Russell, R.M. and Krajewski, L.J. (1992) Coordinated replenishments from a common supplier. *Decision Sciences*, 610–32.

TRAINING AIDS: VIDEOTAPES

Penn State Audio-Visual Services, tel. (800) 826-0312:

Business Logistics Management Series: Inventory in the Logistics System

Business Logistics Management Series: Adapting the Basic EOQ Model for Logistical Decisions

Business Logistics Management Series: Techniques of Inventory Management
Business Logistics Management Series: The Basic EOQ Model

Tompkins Associates, tel. (919) 876-3667:

Manufacturing: Making a Difference

CHAPTER 6

Facilities planning

6.1 INTRODUCTION

Facilities planning addresses two major logistics decisions: facilities location and facilities layout. The first one is concerned with the selection of the best location(s) for establishing facilities based on cost or other criteria. A location may be a country, city or a specific location within a city. Facilities at the macro level may include warehouses, plants, stores, cross-docking facilities, terminals and customer sites. Micro level facilities are smaller components of a larger facility. For example, a factory may consist of a machine shop, paint shop and a welding shop. A machine shop may have milling machines, grinding machines and drilling machines. The second facilities planning decision is concerned with the optimal arrangement and layout of smaller facilities within a major facility in order to minimize material handling and related costs. The layout and location of facilities play a vital role in minimizing the total cost of logistics. The location of facilities has a huge impact on land and construction costs, local taxes and insurance, labor cost and availability and costs of transportation to or from other facilities. The layout of a facility typically has an impact on intrafacility logistics costs such as material handling costs and the cost of material handling equipment.

Much of the earlier historical work on facilities location was based solely on transportation costs. Simple models were suggested by several scholars working in the areas of land-use planning and economics. For instance, Thunen (1875) suggested that the rent for the use of any land to install production facilities cannot exceed the selling price of the products in the market place less the transportation costs. This theory seems to hold good even today when analysts decide on the location of warehouses and plants. Another theory on facilities location based on the type of production process was proposed by Weber in 1909 (cited in Friedrich, 1929). For weight-losing production processes, in which the final product weighs less than the raw material, it is best to locate the plant nearer to the source of raw material than to the market place. For weight-gaining processes it is better to locate the plant near the market place to minimize transportation costs. For processes with no weight gain or loss of raw materials the location may be anywhere between the source of raw materials and the market place. In general, when plants are located close to the source of raw materials inbound transportation costs are less than the outbound transportation costs. When it is located close to the market place outbound transportation costs are smaller than inbound transportation costs. However, this is true only when the transportation costs depend upon distance. In reality, the transportation mode options between plant to market may not be the same as from raw material source to plant. The competition and the carriers providing the service may be different. The transportation rates may vary based on distance, weight and other factors. All these issues complicate the decision process to select the best location. This is further complicated by the fact that there are other costs that have a direct impact on the location besides transportation costs.

The primary objective of facilities layout is to minimize the total cost of material handling. There are also other factors such as not to have certain facilities adjacent to each other (say the paint shop and welding facilities) and forbidding certain areas for certain facilities. This chapter first gives a classification of facilities planning problems. Then, some approaches and decision models to address the facilities location and layout problems are presented. The first part covers single and multiple facilities location under static and dynamic considerations. It also includes popular location models such as the quadratic assignment model for interacting facilities location, the discrete plant location model and the set covering models. The second part focuses on facilities layout. This includes traditional plant layout methodologies and heuristic-procedure-based facilities layout algorithms such as CRAFT and ALDEP.

6.2 CLASSIFICATION OF FACILITIES LOCATION PROBLEMS

A classification of facilities location problems based on the original taxonomy proposed by Francis and White (1974) is presented here. The classification is based on the characteristics of the facilities and the problem characteristics.

6.2.1 Characteristics of facilities

Depending upon the number of new facilities the problem may be of single or multiple facilities. The type of facilities may be viewed as points or areas. At the macro level a plant may be viewed as a point facility. At the detailed level it will be viewed as an area with several other point facilities such as machine tools, storage areas, etc. When the new facilities are viewed as areas then one has to address the facilities

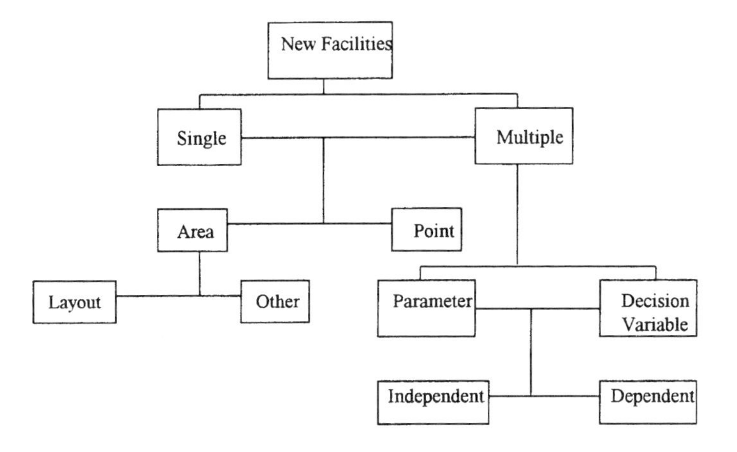

Figure 6.1 Classification based on new facility characteristics.

layout problem for the area. When there are multiple new facilities, the number of new facilities may be a given parameter or a decision variable. A company may want to know the locations for three warehouses or determine the number and locations of warehouses to meet a certain level of service given distances and demands. The new facilities may be dependent [that is, there is interaction among the new facilities (flow of material, information or people)] or independent. A classification of facilities planning problems based on new facility characteristics is shown in Figure 6.1.

6.2.2 Problem characteristics

The facilities planning problem may involve single or multiple dimensions. In a single dimension case the flow between existing and the new facilities occurs along the same line or in single dimension. The problem of locating bus stops along a major commercial street with markets, offices and residential complexes is an example of this. In a multidimensional case, the facilities are not along a single dimension and the flow occurs in more than one dimension. The feasible region for locating facilities may be a discrete set of locations or any location in an open continuous space. In some cases it may be open continuous space with certain forbidden regions. For instance, the available space may be anywhere except locations of existing lakes and fire stations. There may be other constraints such as minimum distance between certain facilities or that new facility 1 and new facility 4, say, have to be close to each other, and so on. The problem may be static or dynamic in nature. In a static situation, the decisions are made for a certain time period and are assumed to be valid for an extended period of time. In a dynamic problem, the decisions may have to be made considering changes in inputs over time. The changes may occur in product flow, facility costs and constraints. Under a static situation, the location and arrangement of facilities may remain the same for an extended period of time. Under dynamic facilities planning the location and/or layout may change every so often, dictated by the dynamic nature of the inputs.

Two other important problem characteristics are the distance function and objective function or evaluation criteria. The type of distance function used for the decision to locate or layout new facilities influences the problem formulation and solution procedures. Most commonly used distance functions are rectilinear and Euclidean. There are other distance functions which are less amenable to derivations and calculations. The objective function may be qualitative or quantitative in nature. The quantitative criteria may be in terms of time, cost or distance. The criteria may be to minimize the total cost or minimize the maximum cost, or some other. A classification of facilities planning problems based on problem characteristics is shown in Figure 6.2.

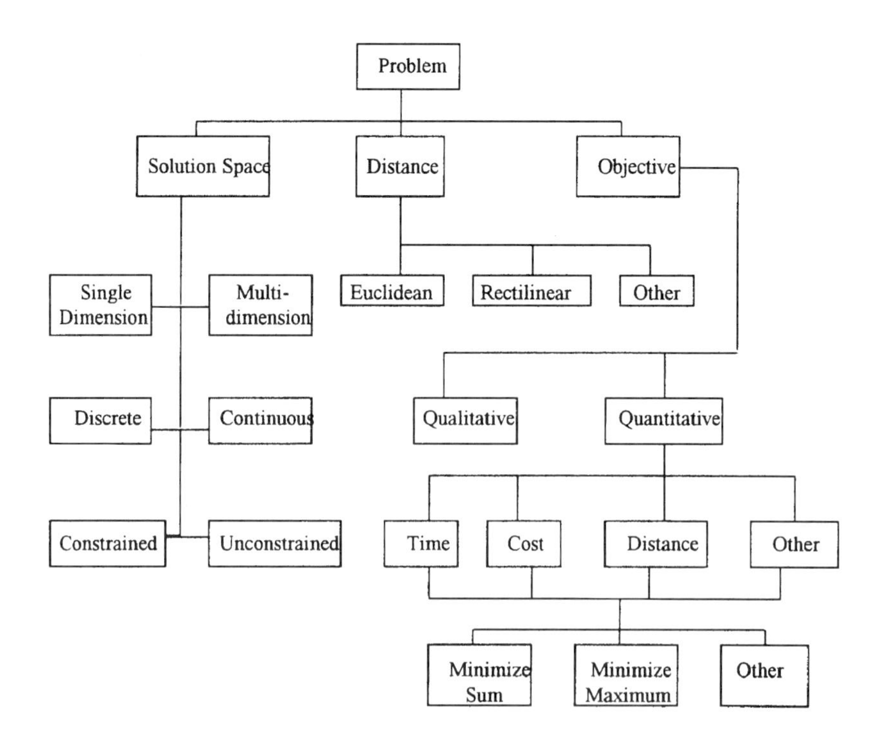

Figure 6.2 Classification based on problem characteristics.

6.3 FACILITIES LOCATION

The primary objective of determining the best location of new facilities is to minimize the total cost of operation. The key cost elements to be considered while addressing facility location problems are transportation cost, operating cost, the fixed cost of locating a facility and the cost of relocating a facility. Transportation costs are based on distance, mode and volume of traffic. Most models assume linear transportation costs. It is also possible to model non-linear costs using a piecewise linear approximation. Operating costs are based on recurring expenses such as labor, utilities and other overheads. The fixed cost of locating a facility is typically the annualized cost of initial investment, which includes the cost of land, equipment and building. Relocation costs include the cost of moving equipment and machinery to the new location, the initial cost incurred at the new location and the cost of closing down the facility at the old location. There are also certain restrictions while locating new facilities. These may include the number and availability of locations, capacity at the sources (plants or warehouses), demand at the sinks (warehouses or customers), total capital available for building new facilities, etc. The complexity of the facility location models depends upon the type and functional form of costs considered and the nature of restrictions imposed.

In this section, several commonly used facility location models are presented. The objective function in almost all cases is based on transportation cost, which is a function of distance and transportation time. Procedures to address single-facility location problems under rectilinear and Euclidean distance functions are presented, followed by a couple of methodologies to extend the single-facility procedure to solve multifacility location problems. A procedure based on a dynamic programming approach is also presented to solve the dynamic or multiperiod location problems. A quadratic programming formulation to model the interaction between new and existing facilities is outlined. Other popular models such as discrete plant location and set covering formulations and their applications are also discussed.

6.3.1 Single-facility location

The problem of locating a single new facility is the simplest of all facility location problems. This problem is addressed here by considering open and continuous solution space and transportation cost. The problem is defined as follows: given the locations of the existing facilities and the volume of traffic between the existing facilities and the new facility, determine the optimal location that will minimize the total cost of transportation, where transportation cost is defined as distance multiplied by volume. Locating a new warehouse with respect to existing plants and markets, building a new store to serve current markets, locating a machine tool in a plant and constructing a regional airport to serve a cluster of towns are some of the examples of single-facility location problems. Here a median-based procedure is used to solve the rectilinear location problems and a center-of-gravity-based procedure is used for solving Euclidean location problems (Francis and White, 1974).

Rectilinear location model

This model is more appropriate when the solution space is a plant, warehouse or a city where the physical movement occurs in a rectilinear manner because of the grid structure of the route system. The rectilinear distance between an existing facility A with coordinates (x, y) and the new facility P with coordinates (a, b), $d(\mathrm{A}, \mathrm{P})$, is defined as follows:

$$d(\mathrm{A}, \mathrm{P}) = |x - a| + |y - b| \tag{6.1}$$

When there are m existing facilities $(\mathrm{A}_1, \mathrm{A}_2, \ldots, \mathrm{A}_m)$ with each having a flow of w_j with the new facility, the problem of new facility location to minimize the total movement can be represented as follows:

$$\text{minimize} \sum_{j=1}^{m} w_j(|x_j - a| + |y_j - b|) \tag{6.2}$$

Equation (6.2) can be rewritten in the following equivalent form [equation (6.3)] which decomposes the problem into two separate minimization problems defined by equations (6.4) and (6.5).

$$\text{minimize} \sum_{j=1}^{m} w_j|x_j - a| + \text{minimize} \sum_{j=1}^{m} w_j|y_j - b| \tag{6.3}$$

$$\text{minimize}\, f(x) = \sum_{j=1}^{m} w_j|x_j - a| \tag{6.4}$$

$$\text{minimize}\, f(y) = \sum_{j=1}^{m} w_j|y_j - b| \tag{6.5}$$

The optimum solution to equations (6.4) and (6.5) satisfy the following two properties (Francis and White, 1974).

1. The x coordinate of the new facility will be the same as the x coordinate of one of the existing facilities. This is true for the y coordinate of the new facility as well. However, the (x, y) coordinates of the new facility may not coincide with the (x, y) coordinates of the same existing facility.

EXAMPLE 6.1

The existing locations of a lathe, drilling machine and milling machine within a production facility are (5, 2), (8, 3) and (6, 5). A new boring machine is to be installed in the facility to meet the changing marketing requirements. The daily volume of flow anticipated between the existing facilities and the boring machine are 50, 25 and 30, respectively. Assuming rectilinear material movement determine the optimal location for the boring machine that will minimize the total distance traveled.

The optimal x-coordinate can be found by listing the facilities in the ascending order of their x coordinate values and selecting the median coordinate value based on flow. This is illustrated in the following table.

Existing facility	*x coordinate value*	*Flow*	*Cumulative flow*
Lathe	5	50	50
Milling machine	6	30	80 (median)
Drilling machine	8	25	105

The median x-coordinate value is 6. The steps in determining the median y-coordinate value are shown in the following table.

Existing facility	*y coordinate value*	*Flow*	*Cumulative flow*
Lathe	2	50	50
Drilling machine	3	25	75 (median)
Milling machine	5	30	105

The median y-coordinate value is 3. Hence, the optimal location for the new facility is (6, 3).

2. The optimum location for the x coordinate (y coordinate) for the new facility is a median location for which no more than one half of the traffic flow is to the 'left' ('below' for the y coordinate) of the location of the new facility and no more than one half is to the 'right' ('above') of the new facility.

An illustration is provided in Example 6.1.

Euclidean location model

This model is more appropriate when the solution space is a region, country or a city where the physical movement occurs along the most direct routes typically represented by modified Euclidean distances.

The modified Euclidean distance, $d(\mathrm{A}, \mathrm{P})$, between an existing facility A with coordinates (x, y) and the new facility P with coordinates (a, b) is defined as follows:

$$d(\mathrm{A}, \mathrm{P}) = K[(x - a)^2 + (y - b)^2]^{1/2} \tag{6.6}$$

K in the above equation is a scaling factor to convert the Euclidean distance to actual distance and it depends upon the country or region under study. When there are m existing facilities $(\mathrm{A}_1, \mathrm{A}_2, \ldots, \mathrm{A}_m)$ each having a flow of w_j with the new facility, the problem of new facility location to minimize the total movement cost can be represented as follows:

$$\text{minimize} \sum_{j=1}^{m} K r_j w_j [(x_j - a)^2 + (y_j - b)^2]^{1/2} \tag{6.7}$$

In the above equation, r_j is the unit transportation cost to location j. The minimization problem represented by equation (6.7) is an unconstrained minimization problem and can be solved by means of simple derivative calculus (Winston, 1987). Taking partial derivatives with respect to a and b, setting them equal to zero and solving for a and b one obtains the following:

$$a = \left(\sum \frac{x_j r_j w_j}{d_j}\right)\left(\sum \frac{r_j w_j}{d_j}\right)^{-1} \tag{6.8}$$

$$b = \left(\sum \frac{y_j r_j w_j}{d_j}\right)\left(\sum \frac{r_j w_j}{d_j}\right)^{-1} \tag{6.9}$$

Note d_j in the above equations represents the distance between the new facility location and the existing facility j and can be computed by using equation (6.6). Since the location of the new facility is not known, the following iterative procedure is used to solve this problem. The iterative procedure is guaranteed to converge to the optimal solution. The procedure is also known as the center-of-gravity approach since equations (6.8) and (6.9) are similar to the equations used for calculating the center-of-gravity in mechanics.

Step 1: determine (a, b) initially using the following equations:

$$a = \left(\sum x_j r_j w_j\right)\left(\sum r_j w_j\right)^{-1} \tag{6.10}$$

$$b = \left(\sum y_j r_j w_j\right)\left(\sum r_j w_j\right)^{-1} \tag{6.11}$$

Step 2: with (a, b) as the new facility location compute d_j by using equation (6.6).

Step 3: solve for (a, b) by using equations (6.8) and (6.9).
Step 4: check if the coordinates (a, b) have changed by more than a specified small value. If yes, go to step 2; if no, stop, the optimal solution has been found.

An illustration is provided in Example 6.2.

EXAMPLE 6.2

ABC Corporation is planning to build a new distribution center to channel production from its two plants to the eight existing retail outlets. The locations of the plants and retail outlets and the expected monthly flow of parts between the plants and retail outlets and the distribution center are given in the table below.

Existing facility	*Location*	*Anticipated daily flow*
Plants:		
1	(5, 45)	8000
2	(50, 0)	9000
Retail outlets:		
1	(20, 50)	4000
2	(40, 45)	1500
3	(0, 35)	500
4	(40, 30)	400
5	(10, 25)	2000
6	(40, 20)	700
7	(30, 15)	3000
8	(10, 10)	1600

The locations are mileage points on the grid. For instance, retail outlets 4 and 6 are only 10 miles apart – outlet 4 is 10 miles north of outlet 6. Assuming a transportation cost of $0.75 per mile and Euclidean movement determine the optimal location for the distribution center that will minimize the total distance traveled.

The problem can be solved by using a straightforward application of the iterative procedure described above. Step 1 results in $a = 26$ and $b = 25.5$. With this as the initial location for the distribution center, steps 2 and 3 are performed, which results in $a = 25$ and $b = 25.5$. Steps 2 and 3 are then repeated with use of the distribution center location of (25, 25.5), resulting in a new location (25, 25.7). The relative change is very close. Hence, the final center-of-gravity-based location for the distribution center is (2, 25.7).

Table 6.1 Important site selection factors (Ghosh and McLafferty, 1987)

Category	*Factors*
Cost structure	Land acquisition cost
	Construction cost
	Taxes, insurance and others
Legal	Covenants and zoning
	Lease clauses
	Other local merchant regulations
Demographics	Population base
	Income potential
	Labor supply
Traffic	Type and volume of traffic
	Transportation modes
	Access to terminals or ports
Retail structure	Competitors
	Type of neighboring stores
Site characteristics	Parking availability
	Building condition (if existing)
	Conditions concerning the visibility of the site from the main street.

Weighted factor analysis

This method can be used to consider quantitative as well as qualitative factors. It requires a list of candidate sites. The first step in weighted factor analysis is to identify the important factors or criteria to be considered in selecting a site and the weights or relative importance of the criteria. The factors identified as important for site selection by Ghosh and McLafferty (1987) may prove to be a good starting point for this step. Table 6.1 provides a sample list of important site selection factors.

The second step is to rate each site against all the factors on a scale of 1 to 10. The last step is to compute a weighted rating for each site and select the site with the highest weighted rating. If more than one facility (say p facilities) are to be located, then simply choose the locations that correspond to the top p ratings. The weighted rating can be computed using the following equation:

$$v(j) = \sum_{i=1}^{n} w(i) \times s(i, j) \tag{6.12}$$

where

$v(j)$ is the weighted rating for site j;
$w(i)$ is the weight for factor i;
$s(i, j)$ is the score for site j on factor i;
n is the number of factors.

An illustration is provided in Example 6.3.

EXAMPLE 6.3

A clothing retail chain has short-listed two locations as potential candidate sites for the new store. The evaluation of the two sites with respect to a few critical factors is given below (higher scores indicate a better site).

Factor (weight)	*Site 1*	*Site 2*
Population base (8)	5	7
Income potential (7)	7	5
Availability of mass transit (3)	4	8
Number of competing stores (6)	6	2
Parking space (7)	4	5
Land and construction costs (5)	9	6

Determine the best site for locating the store.

The weighted rating for site 1 is 210 whereas the weighted rating of site 2 is 192. Since higher scores imply better site, the store should be located at site 1.

6.3.2 Multifacility location

The problem of locating more than one new facility (n new facilities) with m existing facilities is trivial when $n = m$; simply locate one new facility at each of the existing facilities. The problem becomes much more difficult when $m \geq n$. This problem can be addressed by considering open and continuous solution space and transportation cost. Locating new warehouses to serve the existing plants and markets, building break-bulk terminals for an LTL (less-than-truckload) network, locating storage areas in a production facility and constructing fire stations in a city are some of the examples of multiple facility location problems. A couple of approaches are presented to extend the median and center-of-gravity based procedures to solve the rectilinear and Euclidean location problems. The first approach clusters the existing facilities into prespecified groups and then locates one new facility to serve each group. The second approach is exhaustive in nature. Each of the $(n-1)$ new facilities is assigned to one of the $(m-1)$ existing facilities. The optimal location for the one remaining new facility is found based on the remaining $(m-n+1)$ existing facilities. The total distance or cost of the above assignment is computed. This is repeated for all possible assignments of the $(n-1)$ new facilities to the $(m-1)$ existing facilities. The corresponding total cost or distance values are computed. The assignment resulting in minimum total distance or cost is selected. Both the approaches can be used even when the number of new facilities is not prespecified and is a decision variable. The above mentioned approaches to solve the multifacility location problems are illustrated in Example 6.4.

EXAMPLE 6.4

A company is interested in building two distribution centers to meet the growing market demand. The customer locations are grouped into four regions. The locations of the regions and the demands at these regions are given below.

Customer region	*Location*	*Volume (cwt)*	*Transportation rate ($/cwt/mile)*
1	(3,8)	5000	0.04
2	(8,2)	7000	0.04
3	(2,5)	3500	0.095
4	(6,4)	3000	0.095

Clustering-based Method

Let m denote the number of customer regions and n denote the number of new distribution centers. The clustering model has two steps. In the first step, the m customer regions are partitioned into n groups based on their distance-based proximity. Single linkage clustering methods discussed in Chapter 2 may be used to group the existing facilities. In the second step, the optimal location for the new facility for each group is determined by using the median or center-of-gravity method. When the number of new facilities is not prespecified, appropriate clustering methods (Romesburg, 1984) can be used to determine the number of groups which will then be equal to the number of new facilities needed.

In this example, $m = 4$ and $n = 2$. First, cluster the existing facilities into two groups by using the single linkage clustering method with distances as similarity coefficients. Here the recti-linear distance metric is used to develop the following distance matrix.

From	*To*			
	1	*2*	*3*	*4*
1	0	11	4	7
2	11	0	9	4
3	4	9	0	5
4	7	4	5	0

Application of the single linkage clustering method yields the following two groups.

Group 1: customer regions 1 and 3;
Group 2: customer regions 2 and 4.

Now we have two, single-facility location problems, one with customer regions 1 and 3 and the other with regions 2 and 4.

Customer region	*Location*	*Volume (cwt)*	*Transportation rate ($/cwt/mile)*
Problem 1:			
1	(3, 8)	5000	0.04
3	(2, 5)	3500	0.095
Problem 2:			
2	(8, 2)	7000	0.04
4	(6, 4)	3000	0.095

The optimal location for the new facility to serve group 1 is at region 3 since it will result in maximum savings in transportation cost. Similarly, the optimal location of the new facility to serve group 2 is at region 4. Hence the new distribution centers must be located in customer regions 3 and 4.

Exhaustive enumeration

For this example, one possible assignment is to assign customer region 1 to new facility 1 and then customer regions 2, 3 and 4 to new facility 2. Under this assignment, the location of new facility 1 will be the same as the location of customer region 1. The location for new facility 2 can be determined using the median or center-of-gravity method. Another possibility is to assign customer region 2 to new facility 1 and existing facilities 1, 3 and 4 to new facility 2. In this case the location of new facility 1 will be the same as the location of customer region 2, and the location for new facility 2 can be determined based on the median or center-of-gravity method. The different assignments possible are given in the following table.

Assignment	*New facility 1*	*New facility 2*
1	Region 1	Regions 2, 3 and 4
2	Region 2	Regions 1, 3 and 4
3	Region 3	Regions 1, 2 and 4
4	Region 4	Regions 1, 2 and 3

There are only four possible assignments since distribution

centers 1 and 2 are identical. The location for the second distribution center is done by means of the median method. The optimal locations for the second distribution center and the transportation costs for the various assignments are given in the following table.

Assignment	*Second facility location*	*Total transportation cost ($)*
1	(6, 2)	3457.50
2	(3, 5)	2072.50*
3	(6, 4)	2520.00
4	(3, 2)	3930.00

* Optimal location (least transportation cost).

The optimal location for the two distribution centers are (8, 2) and (3, 5). Region 2 should be assigned to the first distribution center at (8, 2) and regions 1, 3 and 4 to the second distribution center at (3, 5).

6.3.3 Dynamic facilities location

In several situations, the location decisions are dynamic in nature. This may be because of change in demands from existing facilities, change in the new facility's capacities and new facility operating costs, from period to period. In a dynamic situation the new facility may have to be relocated to meet the changing cost structure and other requirements. This may result in significant costs which include cost of closing a facility at the old location, cost of moving equipment and parts from the old to the new location and the initial purchase and set-up cost at the new location. In certain cases relocation may not be possible because of contractual or legal reasons; in this case, the best approach is to average the demand and capacity data for the future periods, convert the costs to present values and then solve the problem as a single-period model. When relocation is possible and the relocation costs are negligible simply solve the problem for each period separately. When the relocation costs are significant the problem has to be solved by considering the location cost in each period and the relocation cost from period to period. The dynamic facilities location problem is very similar to the problem of selecting the machine tool for each operation of a part considering sequence-dependent set-up costs (Kasilingam, 1996). The processing cost for an operation on a machine tool corresponds to the location cost for a site in a period. The sequence-dependent set-up cost corresponds to the cost of relocation from one period to another. Here an approach is presented based on the concept of backward dynamic programming to solve the dynamic location of a single new facility when the relocation costs are significant.

The notation to be used is as follows:

S is the set of available sites; these sites are the optimal locations obtained by solving the problem separately for each period;
N is the total number of periods in the planning horizon;
$C^{\ell}(K_n)$ is the cost of locating the new facility in location K in period n;
$C^{r}(A_{n-1}, K_n)$ is the cost of relocating the facility from location A in period $n-1$ to location K in period n;
$C_n^{tot}(A_{n-1}, K_n)$ is the total cost in period n for relocating the facility from location A in period $n-1$ to location K in period n;
$L_n^*(A_{n-1})$ is the best location option in period n if the location was A in period $n-1$

Dynamic programming model

The recursive equations to solve the problem for periods 2 through n can now be written as follows:

$$C_n^{tot}(A_{n-1}, K_n) = C^{\ell}(K_n) + C^{r}(A_{n-1}, K_n) \tag{6.13}$$

$$L_n^*(A_{n-1}) = \min_{K \in S}\{C_n^{tot}(A_{n-1}, K_n)\} \tag{6.14}$$

The equation for period 1 is given as

$$L_1^*(K_1) = C^{\ell}(K_1) + L_2^*(K_1) \tag{6.15}$$

The step-by-step procedure required to solve the dynamic location problem is as follows:

Step 1. select the best location for period 2 through n by using equations (6.13) and (6.14);
Step 2: select the best location for period 1 by using equation (6.15);
Step 3: obtain the optimal location–relocation plan by tracing through the results of steps 1 and 2.

An illustration is given in Example 6.5.

6.3.4 Mathematical models for facilities location

Most of the facility location models can be formulated as mathematical models. The formulations may result in linear or non-linear programming models depending upon the type of problem – whether it involves distance or cost functions, interaction between existing and new facilities and the nature of the constraints. The advantage of mathematical modeling is its ability to handle constraints specific to the particular problem in hand. Some of the common constraints in real-life

EXAMPLE 6.5

A facilities planning manager is faced with the problem of selecting a site for locating a machine shop. Forecasts of cost and flow data for the next four periods exhibit a significant amount of variation. Based on a preliminary study three sites have been chosen. The costs of operating the machine at the three different sites for the next four years are as follows:

Site	*Cost ($ thousands), by year*			
	1	*2*	*3*	*4*
1	60	80	60	120
2	40	80	100	100
3	80	20	120	80

The relocation costs are as follows.

Site		*Relocation cost ($ thousands)*
from	*to*	
1	1	0
1	2	2
1	3	1
2	2	0
2	1	2
2	3	1
3	3	0
3	1	2
3	2	1

Determine the optimal location-relocation plan.

The above example can be solved by using the backward dynamic programming procedure. Sample calculations for the four steps are presented below.

First (step 1), the best locations for periods 4, 3 and 2 are computed by means of equations (6.13) and (6.14). Sample calculations for period 4 with location 1 as the candidate for period 3 are as follows:

$$C_4^{\text{tot}}(1, 1) = C^{\text{r}}(1_3, 1_4) + C^{\ell}(1_4) = 0 + (20 \times 6) = 120$$
$$C_4^{\text{tot}}(1, 2) = C^{\text{r}}(1_3, 2_4) + C^{\ell}(2_4) = 2 + (20 \times 5) = 102$$
$$C_4^{\text{tot}}(1, 3) = C^{\text{r}}(1_3, 3_4) + C^{\ell}(3_4) = 1 + (20 \times 4) = 81$$

Hence, with location 1 as the candidate for period 3 the best location for period 4 is location 3 [$C_4^{\text{tot}}(1, 3) = 81$]. Similar calculations with other locations as the candidates for period 3 lead to the best locations for period 4 and are summarized below, along with the best locations for periods 3 and 2.

Location for Period 3	*Best Location for Period 4*	*Total Cost*
1	3	81
2	3	81
3	3	80
Location for Period 2	*Best Location for Period 3*	*Total Cost*
1	1	141
2	1	143
3	1	143
Location for Period 1	*Best Location for Period 2*	*Total Cost*
1	3	164
2	3	164
3	3	163

Second (step 2), the best location for period 1 is computed by using equation (6.15):

$$L_1^*(1_1) = 224$$
$$L_1^*(2_1) = 204$$
$$L_1^*(3_1) = 243$$

The optimal solution is obtained as follows. The best location for period 1 is location 2 [$L_1^*(2_1) = 204$]. Given that the best location for period 1 is location 2, the best location for period 2 is location 3. Given this, the best location for period 3 is location 1. Based on this, the best location for period 4 is location 3. The optimal location-relocation plan is as follows:

Period	1	2	3	4
Location	2	3	1	3

location problems include capacity restrictions on the new facilities, restrictions on serving an existing facility, such as new facility 1 not being able to serve old facility 3, a limit on the number of new facilities, the total budget available to build new facilities and the compatibility between new facilities, such as if new facility 3 is built then new facility 6 cannot be built, or if new facility 1 is built then new facility 4 should also be built. In this section, the following mathematical models for facilities location are presented:

- simple formulation of a location model;
- simple plant location model;
- set covering models;
- quadratic programming model.

The first is a very simple and fundamental formulation of the multi-facility, discrete solution space location problem. The second is a mathematical model that represents the location of new facilities to serve the existing facilities considering fixed and variable costs of location. The third model focuses on modeling a location problem to provide adequate coverage with a minimum number of facilities or facilities cost. The fourth model presents a non-linear programming formulation of the facility location problem with interaction. The models presented here may be modified to represent some of the real-life constraints discussed before. Several models and procedures are available in the literature to solve a variety of location problems that arise from several industries. In fact, the literature on location is probably among the top five categories of research areas in industrial engineering and operations research in terms of depth and breadth of coverage.

Multifacility location formulation

In this formulation, the available sites for locating the new facilities are known and fixed. The number of new facilities to be located and the material movement between the new and existing facilities are known. The model assumes no interaction between the new facilities. The notation is as follows:

p is the number of available sites;
m is the number of existing facilities;
n is the number of new facilities;
$X_{jk} = \begin{cases} 1, & \text{if new facility } j \text{ is located in site } k, \\ 0, & \text{otherwise;} \end{cases}$
d_{kl} is the distance between site k and existing facility l;
v_{jl} is the flow between new facility j and existing facility l;
c_{jk} is the cost of locating new facility j at site k, where $c_{jk} = \sum_l d_{kl} v_{jl}$.

The model is formulated as follows:

$$\text{minimize } z = \sum_j \sum_k c_{jk} X_{jk} \tag{6.16}$$

subject to

$$\sum_j X_{jk} \leq 1, \forall\, k$$

$$\sum_{k} X_{jk} = 1, \forall j$$

$$X_{kj}\{0, 1\}, \forall k, j$$

The objective function [equation (6.16)] minimizes the total cost of locating all the new facilities. The first constraint ensures that at most only one facility is located at each site. The second constraint ensures that each new facility is located at only one site. The final constraint represents the binary nature of the decision variables. The model assumes that $p \geq m$. When $m = p$, the first constraint will be an equality. In this case the formulation becomes the same as the popular assignment problem and can be solved by using the Hungarian algorithm (Fabrycky *et al.*, 1972).

Discrete plant location model

This is one of the most popular facility location problems modeled and is analysed extensively in the location literature. The problem focuses on determining the required number, locations and sizes of plants or warehouses to meet the demands of customers at various locations. One of the formulations of discrete plant location problem (Efroymson and Ray, 1966) is presented below. The notation is as follows:

X_{kj} is the fraction of customer k's demand supplied by a plant (warehouse) built at location j;

$Y_j = \begin{cases} 1 & \text{if a plant is built at location } j, \\ 0 & \text{otherwise;} \end{cases}$

c_{kj} is the cost of meeting all the demand of customer k from the plant at j;

f_j are the fixed costs associated with building a plant at location j;

n is the number of new facilities.

The model is formulated as follows:

$$\text{minimize } z = \sum_{k} \sum_{j} c_{kj} X_{kj} + \sum_{j} f_j Y_j \tag{6.17}$$

subject to

$$\sum_{j} X_{kj} = 1, \forall k$$

$$\sum_{k} X_{kj} \leq n Y_j, \forall j$$

$$Y_j \in (0, 1), \forall j$$

$$X_{ik} \geq 0, \forall k, j$$

The objective function [equation (6.17)] minimizes the sum of the cost of locating plants and the cost of meeting the demand from customers. The first term on the right-hand side represents the total variable costs of meeting customer demand, and the second term on the right-hand side represents the fixed cost associated with building plants. The fixed costs include the annualized cost of land acquisition and plant construction and operating costs. The first constraint ensures that the total demand of each customer must be met by supply from some plants. The second constraint ensures that a plant can meet the demand (or non-zero fraction of demand) of some or all customers only if it is built. Another way to look at this constraint is that it ensures that a plant is built only if it is assigned to meet the demand (or a non-zero fraction of demand) of all or some customers. The final two constraints represent the binary and non-negativity restrictions of the decision variables, respectively.

Covering models

An important area of facilities location addresses the issue of coverage to customers when locating facilities. The term 'coverage' may indicate service level or response time. In the context of locating fire stations, a location is said to cover a certain residential area if it is within a certain travel time or distance from that area. In the context of locating service depots, a location is said to cover a customer region if the time for the repair truck to get to the customer site is within an acceptable service level. Hence, coverage is a binary variable based on the service criteria established for a location problem. The main objective is to minimize the number of facilities required to provide coverage to all the customers. If the facility cost varies by location then it may be appropriate to minimize the total cost of locating facilities. Two formulations are presented that represent each of the above objectives.

Total cover problem

This is a class of set covering problem where the objective is to minimize the number of facilities required to provide coverage to all the customers (Toregas *et al.*, 1971). Examples include determining the locations for water fountains in a building or identifying the locations for tool cribs in a large machine shop. The following formulation concerns the total cover problem:

$$\text{minimize } z = \sum_j Y_j \tag{6.18}$$

subject to

$$\sum_j a_{jk} Y_j \geq 1, \forall\, k$$

$$Y_j \in (0, 1), \forall j$$

Y_j is defined as before in the discrete plant location model already presented in this section. The parameter a_{jk} is equal to 1 if customer k is covered by location j. The objective function [equation (6.18)] minimizes the number of facilities required. The first constraint ensures that each customer will be served by at least one plant. The second constraint represents the binary nature of the decision variables.

Set covering problem

This is a variant of the total cover problem in which the objective is to minimize the cost of facilities required to provide coverage to all the customers. The objective function implies that the costs of facilities at different locations are different. Examples include determining the locations of schools to serve different parts of the city or identifying the locations for distribution centers to meet customer demands. The following formulation concerns the set covering problem (Shannon and Ignizio, 1970):

$$\text{minimize } z = \sum_j f_j Y_j \tag{6.19}$$

subject to

$$\sum_j a_{jk} Y_j \geq 1, \forall k$$

$$Y_j \in (0, 1), \forall j$$

Y_j and f_j are defined as before in the discrete plant location model and a_{jk} is defined as before in the total cover problem. The objective function [equation (6.19)] minimizes the total cost of facilities required. The first constraint ensures that each customer will be served by at least one plant. The second constraint represents the binary nature of the decision variables.

In certain cases, budgetary restrictions may be imposed to limit the number of facilities that can be built. In this case, the objective may be to cover the maximum number of customers possible. Problems of these nature are represented by 'partial cover' models (ReVelle and Swain, 1970).

Quadratic programming model

In this formulation the available sites for locating the new facilities are known and fixed. The number of new facilities to be located and the material movement between the new and existing facilities are known. Further, there is also interaction among the new facilities in terms of material flow. The formulation is based on the notation defined in the multifacility location model, except the cost of location, c_{jkqr}, is defined as the cost of locating new facility j at site k and new facility q at site r. This cost term includes the cost of movement between the new facilities j and q, between the new facility j and the existing facilities, and between the new facility q and the existing facilities. The cost of movement is a function of the distance and the volume of flow:

$$\text{minimize } z = \tfrac{1}{2}\sum_j \sum_k \sum_q \sum_r c_{jkqr} X_{jk} X_{qr} \tag{6.20}$$

subject to

$$\sum_j X_{jk} \leq 1, \forall\, k$$

$$\sum_k X_{jk} = 1, \forall\, j$$

$$X_{kj}\{0, 1\}, \forall\, k, j$$

The objective function [equation (6.20)] minimizes the total transportation cost related to locating all the new facilities. The first constraint ensures that at most only one facility is located at each site. The second constraint ensures that each new facility is located at only one site. The model assumes that $p \geq m$. When $m = p$, the first constraint will be an equality. The final constraint represents the binary nature of the decision variables.

When there are no existing facilities, the traffic flow is based only on the interaction between the new facilities – the most common example for this is the facilities layout problem. The following heuristic procedure may be used if there is no interaction between the new and the existing facilities (Tompkins and White, 1984):

Step 1: sort the flow values in ascending order and sort the distance values in descending order;
Step 2: assign the facility pair with the largest flow to the site pair with the smallest distance;
Step 3: continue step 2 by assigning the facility pair with the next largest flow to the site pair with the next smallest distance;
Step 4: continue step 3 until a feasible assignment is made.

An illustration is given in Example 6.6.

Location-allocation model

This model is useful to determine the number of new facilities and their locations and the allocation or assignment of new facilities to serve the existing facilities. A common application of the model is the design of the distribution network which involves the determination of warehouse locations and the assignment of warehouses to serve stores or to receive from plants. The model can also be used for selecting the

EXAMPLE 6.6

Four sites (1–4) are available within a plant to locate the four new machines that are on order. The distances between the sites are given in the following table:

Distance (feet) from site:	*Distance (feet) to site:*			
	1	*2*	*3*	*4*
1	0	8	10	2
2	8	0	4	7
3	10	4	0	9
4	2	7	9	0

The anticipated flow of parts between the four new machines (A–D) is given in the following table:

Flow of parts (units) from machine:	*Flow of parts (units) to machine:*			
	A	*B*	*C*	*D*
A	0	2	8	3
B	2	0	4	9
C	8	4	0	5
D	3	9	5	0

Use the heuristic procedure discussed above to determine the best location for the machines.

Step 1: the flow values in the descending order are (9, 8, 5, 4, 3, 2) and the distance values in the ascending order are (2, 4, 7, 8, 9, 10).

Step 2: the machine pair with the largest flow value, of 9, is (B, D). The site pair with the least distance, of 2, is (1, 4). Hence, assign machine B to site 1 and machine D to site 4, or machine B to site 4 and machine D to site 1.

Step 3: match machine pair (A, C), with a flow value of 8, to the site pair (2, 3), with a distance value of 4. Hence, assign machine A to site 2 and machine C to site 3 or machine A to site 3 and machine C to site 2.

Step 4: all assignments are feasible.

From the options available for assignment, we have four alternative solutions. One of the solutions is as follows:

Machine	A	B	C	D
Site	3	1	2	4

The corresponding total material handling distance is 164 feet if the material handling lot size is assumed to be equal to 1.

location of bus depots and assigning buses to the depots in an urban bus transit system. A mathematical formulation of the location-allocation model is presented below (Sherali and Adams, 1984). The notation is as follows:

c_{kjl} is the cost of meeting all the demand of customer k from plant j built at location l;
d_k is the annual demand of customer k;
f_{jl} is the annualized cost of building a plant j at location l;
s_j is the capacity of plant j.

The decision variables are as follows:

X_{kj} is the number of units transported from plant j to customer k;
$Y_{jl} = \begin{cases} 1 & \text{if a plant } j \text{ is built at location } l, \\ 0 & \text{otherwise.} \end{cases}$

The model may be formulated as:

$$\text{minimize } z = \sum_j \sum_l f_{jl} Y_{jl} + \sum_j \sum_k \sum_l C_{kjl} Y_{jl} X_{kj} \qquad (16.21)$$

subject to

$$\sum_l Y_{jl} = 1, \forall j$$

$$\sum_j Y_{jl} = 1, \forall l$$

$$\sum_k X_{kj} = s_j, \forall j$$

$$\sum_j X_{kj} = d_k, \forall k$$

$$Y_{jl} \in (0, 1), \forall j, l$$

$$X_{kj} \geq 0, \forall k, j$$

The objective function [equation (6.21)] minimizes the sum of the cost of locating plants and the cost of meeting the demand from customers. The first term on the right-hand side represents the fixed cost associated with building plants, and the second term on the right-hand side represents the total variable costs of meeting customer demand. The fixed costs include annualized cost of land acquisition and plant construction and operating costs. The first constraint ensures that a plant is built at only one location. The second constraint ensures that each location has only one plant. The third constraint represents the capacities of plants, and the fourth constraint represents the demands of customers. The last two constraints represent the binary and non-negativity restrictions of the decision variables, respectively.

The location-allocation problem can be solved heuristically by using an iterative procedure (Cooper, 1964). The procedure starts with the initial selection of locations for the new facilities (the Y_{jl}s). Based on the initial locations, the allocation problem is solved to obtain the X_{kj}s. Now, using the allocations as given, the location problem is solved again. The procedure is repeated until convergence is reached. The test for convergence is usually the closeness of the objective function values for the successive iterations. The procedure may also be terminated based on a certain maximum number of iterations.

6.4 FACILITIES LAYOUT

The primary objective of determining the best arrangement or layout of facilities within a plant or warehouse is to minimize the total cost of material handling. The material handling cost depends upon the distance between the facilities, the type of material handling equipment and the flow of material between the facilities. The type of distance function may depend upon the type of material handling equipment: an automated guided vehicle moves along a rectilinear path and conveyors transport material along the shortest path in a straight line defined by Euclidean distance. The type of material handling equipment depends upon the type of parts or products to be transported, the amount of material handling flexibility required and the cost of the systems – both the initial and the operating costs. The flow of material depends upon the process plans of the various products manufactured and the production volume. The constraints are the sizes of the different facilities, the compatibility or proximity restrictions on facilities and the structural characteristics of the plant or warehouse: the space available in the southwest corner of the building may not be able to house the mixing department; the welding shop and the painting area cannot be next to each other; or the second floor may be too weak structurally to house the drilling unit. This section first presents some of the commonly used layout types and then offers a few approaches to solve the layout problem.

6.4.1 Common layout types

The following four commonly used layout types are presented below: product layout, process layout, fixed position layout and cellular layout. The type of layout depends upon the production volume, variety, process plans and the weight or size of the products.

Product layout

Under this layout the production facilities are arranged along a line in a serial manner to represent the flow of the product through the various facilities. For instance, if the operation sequence requires a lathe, a milling machine, heat treating and painting then the production line will first have the lathe, then the milling machine followed by the heat treatment shop and the paint shop. This type of layout is more suitable when all products have a similar flow through the system. In other words, product variety should be very limited and production volume must be very high to justify this layout. Good examples of such a product layout are bottling plants, automobile assembly lines and food packaging lines. The advantages of product layout are reduced material handling and inventories and easier control and supervision. On the other hand, the flexibility of the layout to changes in product variety is very low. Any changes in product design may lead to a major investment in creating a new layout. Also, the production rate is determined by the operation that takes the longest time. Balancing of the line is critical to achieve maximum production rate and minimum idle time.

Process layout

In this layout, production machines are grouped based on the function or processes they perform. For instance, all drilling machines are grouped and put in one place and all welding units are placed together in another spot. This layout supports job shop type of production units where product variety is very high and production volume is low. It provides good flexibility and agility to meet market needs and leads to lower investment in facilities. However, it increases material handling, production time and requires careful planning and close supervision. Good examples are companies that manufacture parts to order for several client companies.

Fixed position layout

This layout is appropriate where the product to be manufactured is too heavy and large to move around in a plant to go through a series of facilities. In this layout, the product stays in one location and the resources, such as workers, material and tools, are brought to its location. This is commonly used in heavy construction work such as building a dam, constructing an office and the manufacturing of a ship. In most cases, this type of layout, because of the nature of the product involved, requires the cumbersome movement of materials and equipment.

Cellular layout

This layout stems from the concept of group technology (GT). GT is a philosophy that capitalizes on the similarities in recurring activities. Cellular layout is an application of GT where machines are grouped into cells based on their capabilities to process a family of parts. The first step is to identify products that are similar in terms of processing requirements and then to group them into families. The next step is to pool the machines required to process one or more families into one cell. The objective is to minimize the intercell movement of products. Within a cell, machines have to be arranged optimally to minimize intracell material handling. This type of layout is appropriate where products can be grouped into families. This layout is less attractive when the processing requirements of products differ greatly.

6.4.2 Layout planning methodologies

Three methodologies for developing the best layout where the four common types of layouts may not be applicable or relevant are presented here. The first methodology is the systematic facilities layout process as described by Muther (1966). This process is very similar to any engineering decision process. It involves the generation of alternative layouts and selecting the best one based on some criteria.

The next two methodologies are computer-based approaches: CRAFT and ALDEP. The first falls into the broad category of improvement methods and the second falls into the category of construction methods. CRAFT starts with an initial layout and then makes a sequence of systematic pairwise exchanges of facilities in order to improve the layout. The procedure stops when further exchanges do not result in further reduction in material handling. ALDEP builds the layout from scratch, making both selection and placement decisions. The selection decision determines the next facility to be placed into the plant and the placement decision focuses on where exactly to place it.

Systematic layout planning

The procedure is represented in the form of a flow chart in Figure 6.3. The first step in the procedure is to develop an activity relationship chart and to construct a from–to chart. The activity relationship chart shows the desired closeness ratings between the various departments or shops within a plant and the reasons for the ratings (Figure 6.4). The definitions of the closeness ratings are as follows:

A absolutely necessary;
E especially important;
I important;
O ordinary closeness is okay;
U unimportant;
X undesirable

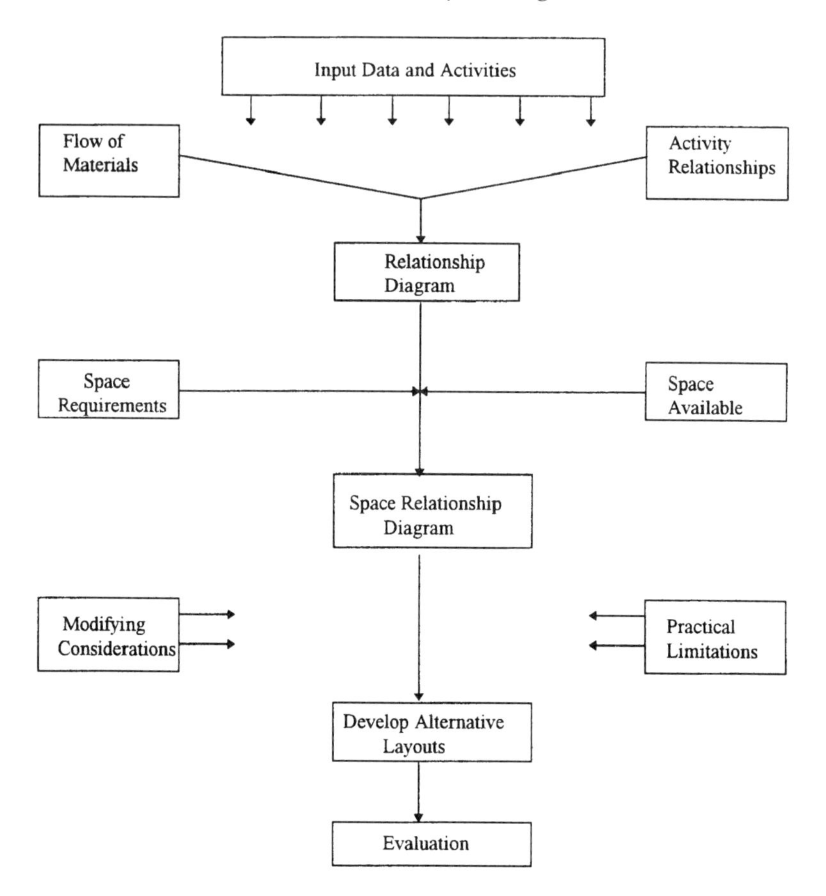

Figure 6.3 Systematic layout planning process.

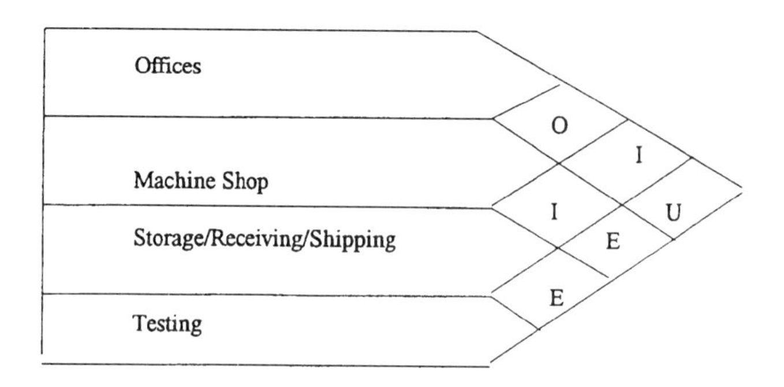

Figure 6.4 Example of an activity relationship chart. Note: typically the box containing the closeness rating is divided into two halves – one containing the rating and the other the reason code for rating.

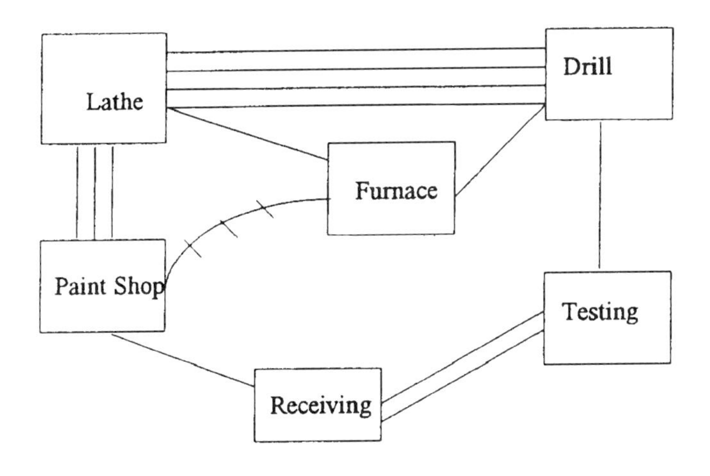

Figure 6.5 Example of a relationship diagram. Closeness ratings: A (absolutely necessary) = four connecting lines; E (especially important) = three lines; I (important) = two lines; O (ordinary closeness is okay) = one line; U (unimportant) = no lines; X (undesirable) = crossed line.

Some of the reasons for assigning a particular closeness rating may be communication, convenience, high material flow, same supervisor and service requirement. A from–to chart shows the flow of material or parts between various shops or departments. Combination of the from–to chart and the activity relationship chart yields the relationship diagram (Figure 6.5). In this diagram each activity or machine or department is represented by an equal sized square. The squares are connected by a certain number of lines that represent the closeness rating.

The required space for each department is determined based on the type of equipment, operator requirements and the material flow and storage needs. The relationship diagram is now modified so that the size of a department is an indication of its spatial requirements. The modified diagram is known as the space relationship diagram (Figure 6.6). The next step is to develop several alternatives based on the space relationship diagram and by considering other practical limitations. Finally, the best alternative is selected based on certain criteria, most commonly a combination of distance and volume of material flow.

CRAFT

As discussed before, CRAFT (computerized relative allocation of facilities technique) (Armour and Buffa, 1963) is an improvement-based

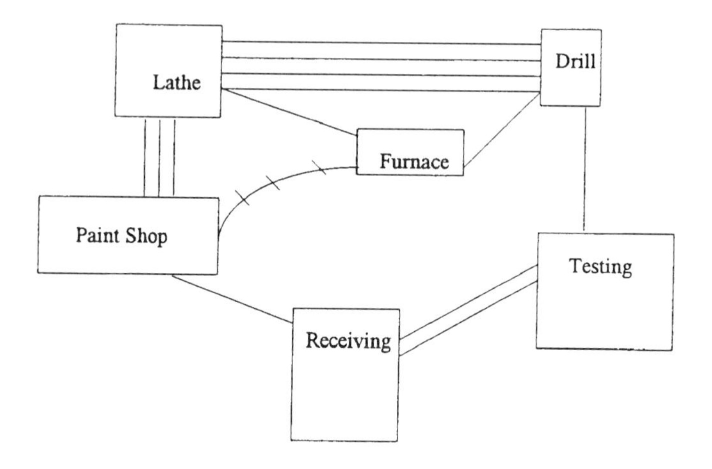

Figure 6.6 Example of a space relationship diagram. For the coding used with respect to the number and type of connecting lines, see Figure 6.5.

method of determining facilities layout. The criteria used for selecting the best layout is transportation cost and it is assumed to be linearly related to distance. The key input requirements are initial layout, flow data in the form of a from–to chart, cost data (cost per unit per unit distance) and the number and location of fixed departments. The initial layout provides information on the number of departments and their respective areas. CRAFT requires all departments to be square or rectangular. If there are any interior voids they have to be represented as dummy departments. Similarly, departments that are fixed, such as restrooms, staircases and aisles, are also represented as dummy departments. Distances are calculated in CRAFT as the rectilinear distance between the centroids of the departments. The algorithm used in CRAFT is as follows.

Step 1: given the initial layout, compute the distance matrix using the centroids of the departments. Compute the corresponding total transportation cost by means of flow, distance and cost matrices.

Step 2: consider interchanging the locations of departments that are of equal area or that have a common border. If no interchange produces a better layout than the present, then stop; otherwise, go to step 3.

Step 3: Select the interchange resulting in maximum savings and go to step 2.

CRAFT allows pairwise, threeway, pairwise followed by threeway, and threeway followed by pairwise interchanges. A threshold value may be specified for making an interchange. For instance, if the maximum saving is less than x dollars, then CRAFT will not make that move. CRAFT does not guarantee optimality since all possible interchanges are not considered.

ALDEP

ALDEP (automated layout design program) (Seehof and Evans, 1967) is a construction-algorithm-based facilities layout program. Construction-based algorithms make selection as well as placement decisions. Selection focuses on the next facility to be placed into the layout, and placement decisions are concerned with where exactly to place that facility in the layout. The criteria used for selecting departments are the closeness ratings. The algorithm used for selection in ALDEP is outlined below.

Step 1: select a department randomly;
Step 2: select a department having a closeness rating A with the selected department. If there is a tie, break it arbitrarily. If no department has a minimum specified closeness rating then choose one randomly.
Step 3: repeat step 2 on the second department selected. Continue until all departments are selected.
Step 4: compute the total score for this layout as the sum of the closeness ratings of all adjacent departments.
Step 5: repeat steps 1 through 4 to generate a prespecified number of alternative layouts (say N). Also, ignore layouts with a total score less than a prespecified value (say Z).

It is to be noted that ALDEP does not take distances into consideration while evaluating the alternative layouts. Closeness ratings are assigned certain numerical values in order to compute the total score: A = 64, E = 16, I = 4, O = 1, U = 0 and X = −1024. The placement decision follows a certain pattern in ALDEP. The first department is placed in the upper left corner and extends downward. The width (known as the sweeping width) is specified by the user. The next department is placed at the end of the previous department and the placement process continues in a zigzag pattern.

6.5 SUMMARY

Facilities planning represents one of the key planning areas within the logistics umbrella. The location or relocation of new facilities and the laying out of facilities within a location are the key logistics decisions that have to be made at the initial stages of planning and designing a logistics system. Transportation cost, one of the major cost components in logistics, is effected by facilities planning decisions. The size and location of the facilities have an impact on inventory costs and the level of customer service. Most of the location problems that arise in real-life today are far more difficult to solve. The difficulty arises from both the problem size and the problem complexity. Typically, mathematical models combined with heuristic procedures are used to solve these problems. Simulation models may be used to evaluate the performance of the system for a given choice of locations of facilities. In most cases the layout and location problems are solved in a hierarchical manner. The two problems may have to be solved jointly if the sites available vary in size and other characteristics. In this case, the layout decisions may depend upon the site selected. A given site may be the best in terms of location costs or interlocation transportation costs plus initial costs but may not be the best in terms of layout costs or intrafacility handling costs. A more comprehensive approach is to solve the problem at the logistics network level considering all other costs as well (inventory costs, warehouse operating costs, etc.). This aspect was discussed in Chapter 3

PROBLEMS

6.1 Discuss the importance of facilities location and layout decisions in the context of logistics and transportation.

6.2 What are the underlying assumptions of the various location models presented in this chapter? Discuss the seriousness of the assumptions.

6.3 Discuss the relationship (sequential or integrated) between location and layout decisions, giving suitable examples.

6.4 A company has two major customers 500 miles apart along a straight highway that runs east–west. The number of service calls from customer A located on the eastern side is 200 units per month and the volume of service calls from customer B located on the west end is 100 units per month. Determine the best location to build a dispatch center for repair technicians along the same highway.

6.5 Develop a linear programming formulation for the above problem and verify your solution.

6.6 Dashboard Corporation is planning to locate its new plant to serve three of its existing customers. The three customers are located along Interstate 95; customer A is in Savannah, GA, customer B is in Jacksonville, FL and customer C is in Miami, FL. The distance between Savannah and Jacksonville is approximately 150 miles, and the distance between Jacksonville and Miami is approximately 350 miles. Jacksonville is north of Miami, and Savannah is north of Jacksonville. Where should the plant be located if Jacksonville has the highest demand? Will the decision change if Miami has the highest demand?

6.7 The (x, y) coordinates, volume of material flow and transportation rates for two stores and three plants are given in the following table.

Site	*Location*	*Volume (units)*	*Transportation rate ($/unit)*
Store 1	(2, 5)	400	4
Store 2	(3, 8)	500	3.5
Plant 1	(6, 4)	600	2
Plant 2	(8, 8)	350	3
Plant 3	(8, 2)	400	2

Material flow is along the shortest path. Determine the optimal location for a warehouse to serve the stores and the plants.

6.8 For the data given in problem 6.7, if the decision is to locate two warehouses determine the center-of-gravity locations.

6.9 Assuming rectilinear material movement, determine the best median location for the data given in problem 6.7

6.10 The distance between five small hospitals that participate in a health care network is as follows.

Distance (miles) from hospital:	*Distance (miles) to hospital:*				
	1	*2*	*3*	*4*	*5*
1	0	28	34	12	8
2	28	0	29	25	18
3	34	29	0	22	32
4	12	25	22	0	10
5	8	18	32	10	0

Two blood banks have to be located to serve all five hospitals. Every hospital must be served by both banks, and a bank can be located only within a hospital. Determine the optimal locations for the banks if the definition of service is less than 30 miles between a bank and a hospital.

6.11 ABC company has plants in four different cities and has warehouses in five different states. Customers are supplied direct from the warehouses. The warehouse managers are responsible for ordering the product from the plants, and products may be ordered from any of the plants. The requirements for the product for June 1997 at the five warehouses are 200, 300, 500, 400 and 400 units, respectively. Develop an optimal or near-optimal procurement and transportation plan given the following information:

	Plant			
	1	*2*	*3*	*4*
Transportation cost ($/unit): warehouse				
1	0.2	0.21	0.23	0.2
2	0.21	0.22	0.22	0.23
3	0.22	0.2	0.21	0.22
4	0.2	0.21	0.2	0.21
5	0.21	0.2	0.2	0.2
Plant capacity (units)	600	400	500	500
Production cost ($/unit)	2	2.1	2.2	2

6.12 Develop a generic mathematical model to address the multifacility, multiperiod discrete location model considering costs of transportation and relocation.

6.13 Develop an analogy between the dynamic facility location problem and intermodal transportation. Note: in an intermodal transportation problem, multiple modes may be available to transport a shipment on any given segment of the route.

6.14 As a logistics engineer you have been assigned the responsibility of laying out the company's receiving and shipping dock. Clearly describe your approach, giving data requirements and procedure, to solve this problem.

6.15 Given the flow cost matrix (flow multiplied by cost per unit flow) and the initial layout below use the CRAFT pairwise exchange technique to obtain an improved layout. Note: all departments are of equal size and measure $4' \times 4'$.

Flow from:	*Flow to:*			
	1	2	3	4
1	0	2	2	12
2	4	0	8	2
3	0	8	0	4
4	4	0	0	0

Initial layout:

P	Q
R	S

REFERENCES

Armour, G.C. and Buffa, E.S. (1963) A heuristic algorithm and simulation approach to relative location of facilities. *Management Science*, 294–309.

Cooper, L. (1964) Heuristic methods for location-allocation problems. *SIAM Review*, **6**, 37–52.

Council of Logistics Management (1994) *Bibliography of Logistics Training Aids*, 2803 Butterfield Road, Suite 380, Oak Brook, IL 60521.

Efroymson, M.A. and Ray, T.L. (1966) A branch-bound algorithm for plant location, *Operations Research*, **14**, 361–8.

Fabrycky, W.J., Ghare, P.M. and Torgerson, P.E. (1972) *Industrial Operations Research*, Prentice-Hall, Englewood Cliffs, NJ.

Francis, R.L. and White, J.A. (1974) *Facilities Layout and Location–An Analytical Approach*, Prentice-Hall, Englewood Cliffs, NJ.

Friedrich, C.J. (1929) *Alfred Weber's Theory of the Location of Industries*, University of Chicago Press, IL.

Ghosh, A. and McLafferty, S.L. (1987) *Location Strategies for Retail and Service Firms*, D.C. Heath, Lexington, MA.

Kasilingam, R.G. (1996) Process plan selection considering sequence-dependent setup cost. *Proceedings of the Fifth Industrial Engineering Research Conference*, Minneapolis, MN, 176–180.

Muther, R. (1961) *Systematic Layout Planning*, Industrial Education Institute, Boston, MA.

ReVelle, C. and Swain, R. (1970) Central facilities location. *Geographical Analysis*, **2**, 30–42.

Romesburg, H.C. (1984) *Cluster Analysis for Researchers*, Lifetime Learning Publications, Belmont, CA.

Seehof, J.M. and Evans, W.O. (1967) Automated Layout DEsign Program. *The Journal of Industrial Engineering*, **18**, 690–5.

Shannon, R.E. and Ignizio, J.P. (1970) A heuristic programming algorithm for warehouse location. *AIIE Transactions*, **2**, 334–9.

Sherali, H.D. and Adams, W.P. (1984) A decomposition algorithm for a discrete location-allocation problem. *Operations Research*, **32**, 878–900.

Thunen, J.H. (1875) *Der Isolierte Staat in Beziehung auf Landwirtschaft und Nationalokonomie*, Schumacher-Zarchlin, Berlin.

Tompkins, J.A. and White, J.A. (1984) *Facilities Planning*, John Wiley, New York.

Toregas, C., Swain, R., ReVelle, C. and Bergman, L. (1971) The location of emergency service facilities. *Operations Research*, **19**, 1363–73.

Winston, W.L. (1987) *Operations Research: Applications and Algorithms*, PWS-KENT Publishing, Boston, MA.

FURTHER READING

Buffa, E.S., Armour, G.C. and Vollmann, T.E. (1964) Allocating facilities with CRAFT. *Harvard Business Review*, **42**, 136–59.

Cook, R.L. (1994) *1994 Bibliography of Logistics Training Aids*. Council of Logistics Management, 2803 Butterfield Road, Suite 380, Oak Brook, IL 60521, pp. 55–78.

Cooper, L. (1963) Location-allocation problems. *Operations Research*, **11**, 331–44.

James, R.W. and Alcorn, P.A. (1991) *A Guide to Facilities Planning*, Prentice-Hall, Englewood Cliffs, NJ.

Moore, J.M. (1962) *Plant Layout and Design*, Macmillan, New York.

TRAINING AIDS: VIDEOTAPES

Penn State Audio-Visual Services, tel. (800) 826-0312:

Business Logistics Management Series: Location Analysis;
Business Logistics Management Series: Location Techniques.

CHAPTER 7

Intrafacility logistics

7.1 INTRODUCTION

Intrafacility logistics is a key area to be addressed within the overall realm of logistics. Intrafacility logistics is concerned primarily with the material handling within a large facility such as a plant or a warehouse. The purpose of material handling is to move raw materials, work-in-process and finished parts as well as tools and supplies from one location to another to facilitate the overall production operation. The handling of material must be performed in a safe, efficient, accurate and timely manner. Material handling is an important aspect of any manufacturing operation. Typically, a part spends almost about half of its manufacturing time in material handling. Intrafacility logistics cost is influenced by layout, material handling equipment, stock locations in the warehouse and operating rules for material handling and storage and retrieval.

There are certain basic principles in material handling. By following these principles (Tompkins and White, 1984) a company can cut costs, increase productivity and reduce accidents and damage. The principles are:

- always transport material in unit loads in a pallet or container;
- avoid partial loads whenever possible;
- move material through the shortest route and distance and use straight-line movement whenever possible;
- minimize terminal time (load, unload and transfer time);
- use gravity to move material;
- carry loads in both directions; in other words, avoid empty travel to the extent possible;
- mechanize material movement;
- maintain orientation of parts during material handling;
- integrate material handling equipment with other equipment in the system and integrate with information flow.

7.2 TYPES OF MATERIAL HANDLING EQUIPMENT

The choices available in terms of different categories of material handling equipment are plenty. Within each category there are different types of equipment. For instance, the category of conveyors includes belt conveyor, roller conveyor and slat conveyor. Material handling equipment can be classified into groups based on the following three criteria: extent of automation, mobility and path trajectory.

Based on the extent of automation, material handling equipment may be grouped into manual, mechanized or automated equipment. Hand trucks, hand trolleys and wheelbarrows are examples of manual equipment. Power trucks such as forklifts and platform trucks and conveyors are examples of mechanized equipment. Automated guided vehicles (AGVs) fall under automated equipment. Material handling equipment may be fixed or mobile. Conveyors and robots are examples of fixed equipment. Forklifts and AGVs fall under the mobile group. Path trajectory may be fixed, variable or programmable. The fixed path equipment includes conveyors and pipelines. Certain types of cranes such as overhead and gantry cranes also fall under fixed path movement. Variable path equipment includes all types of trucks and robots and certain types of cranes. Programmable path equipment includes robots and AGVs. In this section, some additional information on material handling equipment will be presented using path-based classification.

7.2.1 Fixed path equipment

The material handled using fixed path equipment follows the same path between any two stations or machines. The path is fairly rigid over a given period of time and changes to the path usually require additional fixed investment. This type of material handling equipment is suitable for high-volume, low-product-variety production situations and warehousing operations. 'Fixed path' does not mean material is carried along the shortest path or with straight-line movement, although it is desirable to do so. Conveyors are the single largest group of equipment that fall under this category of material handling equipment. There are several different types of conveyors based on the mechanical construction and the type of mechanism used for moving material – belt, screw, slat, roller, chain, overhead trolley and so on. In roller conveyors, the carrying medium consists of a series of tubes or rollers that are perpendicular to the direction of travel. Roller conveyors can be powered or of the gravity type. These are suitable for flat pallets or containers and are often found in warehouses and final packaging lines. Belt conveyors are commonly used for carrying solid bulk material such as minerals or coal. The belt may be flat or troughed. Slat conveyors have individual platforms connected to a moving chain. They operate very much like belt conveyors.

All conveyor types can be classified into the following conveyor systems: single direction, continuous loop and power-and-free. Single direction conveyors carry material one way from origin to destination with no need for returning the pallet or container. Loop conveyors function like a two-directional system where loaded containers move between the load and unload stations and empty containers are returned to the load station. Loop conveyors when used for temporary storage are known as recirculating conveyors. Power-and-free conveyors can be operated in an asynchronous manner allowing independent movement of containers or pallets. This allows for differences in production rates between stations and temporary storage.

7.2.2 Variable path equipment

All types of trucks such as forklifts, hand trucks, power trucks and platform trucks move along a variable path depending upon the requirements. Certain types of cranes that are mounted on trucks also fall under this category. Variable path equipment provides flexibility in terms of material handling but is less efficient in terms of the volume of material handled. Hand trucks have platforms or containers that are mounted on wheels for manual movement of boxes and bulky material. Power trucks may be powered by battery, gasoline or propane. Power trucks have features for moving the load in several directions

7.2.3 Programmable path equipment

AGVs and robots can be programmed to travel in certain paths and directions. In the case of AGVs travel capability is restricted within a limited travel area. The travel area may be the paths constructed within a facility. In other words, AGVs provide more flexibility than does fixed path equipment but less flexibility than does variable path equipment. In addition, AGVs can help to meet changing material handling requirements resulting from changing manufacturing requirements. In the case of robots, the travel area is not restricted to certain paths; robots truly fall under programmable, variable path equipment. In this section, an overview is provided of the different types of AGVs and some of the planning and control problems associated with AGVs.

AGVs are self-propelled, independently operated material handling equipment that operate along defined pathways in the floor. AGVs are typically battery powered, with batteries which may last for approximately 8 to 16 hours. They are independent since one AGV is not controlled by another. The pathways are either wires embedded underneath the floor or reflective paint strips on the floor. AGVs can be broadly classified into the following three categories: driverless trains, pallet trucks and unit load carriers. In the driverless train type, a towing vehicle pulls one or more trailers. This is more appropriate for heavy payloads over long distances with intermediate pick-up and delivery points. In the pallet type, the loads are on pallets and the AGVs have forks to handle the pallets. Typical capacity is around 6000 lb and the vehicle may have to be steered manually off its path to the storage areas. The unit load carriers move loads in units and are equipped with automatic loading and unloading features such as moving belts or powered rollers. Unit load AGVs may be used in assembly situations to transport a product from base component through to final assembly. The efficient and safe operation of AGVs requires three basic systems: a guidance system, a routing system and a traffic control system.

The AGV guidance system may consist of wires or paint strips. The wires are embedded in a channel $\frac{1}{8}''$ wide and $\frac{1}{2}''$ deep or are taped onto the floor. A frequency generator provides the guidance signal by creating a magnetic field along the pathway and the AGV is equipped with two sensors at the bottom on either side of the wire. The paint strip guidance path consists of 1″ reflective fluorescent paint, and the vehicle is equipped with an optical sensor at the bottom. This type of guidance system is useful when there is electrical noise in the system or when burying or placing cables in the floors is not possible. It is hard to maintain the painted strip free from dirt; this may reduce the guidance efficiency. Most AGVs are equipped with certain safety features. For instance, the vehicle may stop if it strays away from the guide path by more than 2″–6″. The vehicle may be brought back to the path if it is within a certain distance known as the 'acquisition distance'.

The physical routing system enables the AGV to go along a selected route. Every intersection where there are branching alternatives is a decision point. Two general methods are used for physical routing. The first one is known as frequency select in which a particular route is followed by the AGV based on its signal frequency. For instance, if the AGV is operating at 5 kHz, then a particular route can be avoided by switching the path frequency to 10 kHz. This method may require several frequency generators. The second method is known as the path switch select method. The concept here is to leave the power on in the selected route and turn off the power in the other routes. This requires independent power control at various route segments.

The traffic control methods prevent collisions between vehicles by controlling the movement of AGVs. This is also known as blocking – preventing any AGV from hitting the one ahead. This may be achieved by on-board vehicle sensing with optical or magnetic proximity sensors. These are not effective at the corners when vehicles are making turns. Another method available for preventing collisions is to use the concept of zone blocking. Under this method, pathways are divided into zones. Each zone can hold a vehicle and will have some extra space for safety. When a zone is occupied, a sensor is activated in the preceding zone. No vehicle is permitted into a zone if it is occupied. In addition, AGVs are also equipped with bumpers, warning lights and bells. Sending AGVs to where they are needed (dispatching) may be done by using on-board control or by using centralized computer control.

7.3 EQUIPMENT SELECTION AND ECONOMIC ANALYSIS

Selection of material handling equipment is often a three-step decision process. In the first step, candidate equipment types are selected based on the functional requirements. The functional requirements that impact the design and selection of the material handling system are as follows:

- characteristics of materials – the characteristics of materials may be categorized into the following (Muther and Haganas, 1969): physical form (solid, liquid, gas), size (length, width, height), weight and volume, shape (long, round or square), damage potential (fragile, brittle, sturdy), hazard rating (explosive, toxic, corrosive) and condition (hot, wet, sticky, dirty);
- the quantity or volume of material flow and flow rate – a large volume of flow will require a dedicated equipment or system; equipment may be shared with other parts with similar characteristics if the volume of flow is small in order to improve utilization and reduce costs; volume and flow rate also determine the capacity and speed of the equipment;
- frequency and scheduling of movements – this relates to the time between movements and the urgency of the movements; the movements may be continuous, in discrete batches or one at a time; the movement may be seasonal, regular or rush orders;
- route factors – this includes the path of travel, distance, elevation, slope and conditions along the route such as temperature, humidity, congestion, turns, and so on.

The second step is to select a set of candidate models of the equipment types selected in step 1. For instance, the candidate equipment types may be pallet trucks and unit load forklifts. Different manufacturers may offer different models of pallet trucks and forklift trucks. Only certain models will meet the functional requirements. The third step is to select the best equipment type or model to meet the material handling requirements based on certain key criteria. The criteria may include costs and/or other tangible, non-economic criteria such as reliability, warranty, maintenance package etc. The first two steps are typically done by engineers or technical managers who are technically knowledgeable about the material handling equipment. The last step may be performed by management analysts or logistics personnel. Expert-system-based approaches may be used for performing steps 1 through 3 (Liang and Dutta, 1988). An expert system contains several rules for equipment selection. The rules are built around the functional requirements. The system matches the requirements to the features of the available equipment types and models in the database and recommends alternatives. The system also considers financial aspects of the selection process. In this section, some methods to perform step 3 will be presented.

7.3.1 Weighted factor analysis

The most commonly used approach for selecting the best among the available alternatives is weighted factor analysis. The first step in weighted factor analysis is to identify the important factors or criteria to be considered in selecting an equipment type or model and the weights or relative importance of the criteria. Table 7.1 provides a sample list of important site selection factors.

The second step is to rate each piece of equipment or model against all the factors on a scale of 1 to 10. The last step is to compute a weighted rating for each piece of equipment or model and select the one with the highest weighted rating. The weighted rating can be computed by using the following equation:

$$v(j) = \sum_{i=1}^{n} w(i) \times s(i, j) \quad (7.1)$$

where

$v(j)$ is the weighted rating for equipment type or model j;
$w(i)$ is the weight for factor i;
$s(i, j)$ is the score for equipment type or model on factor i;
n is the number of factors.

Table 7.1 Important material handling equipment selection factors

Category	*Factors*
Cost structure	Purchasing price
	Operating cost
	Maintenance cost
Capacity	Weight
	Volume
	Type of material that can be handled
Flow path	Direction of travel
	Type of movement (straight, flexible)
	Continuous or discrete flow of material
Operating conditions	Humidity
	Temperature
	Congestion

An example of the weighted factor method is given in section 4.3.2. The analytical hierarchic process (AHP) described in section 4.3.4 can also be used for selecting the best material handling equipment type or model. The AHP model calculates the weights as part of the decision process.

7.3.2 Economic analysis

A variety of analytical methods may be used when only financial factors are considered in selecting the best material handling equipment. The most common methods are the present-worth comparison, annual-cost and internal rate of return methods. These methods are illustrated by Example 7.1.

If the benefits that can be realized by using the material handling equipment can be quantified then payback period analysis and benefit–cost ratio methods can also be used. Other issues to be addressed in selecting the best equipment are unequal lives of candidate equipment, tax impact and depreciation. Any standard text on engineering economy will cover all of these in detail (Riggs, 1982).

7.3.3 Systematic equipment selection procedure

The systematic site selection procedure discussed in Tompkins and White (1984) can be used for selecting material handling equipment as well. The procedure considers critical, subjective and objective factors. The procedure consists of the following steps.

Step 1: define critical factors and identify potential equipment candidates. Critical factors are attributes of the equipment that are absolutely essential to meet the material handling requirements. Any equipment that does not meet the critical factors will not be considered as potential candidates. Table 7.2 illustrates step 1. A value of 1 indicates that the equipment meets the critical factor requirement. Critical factor measure (CFM) is the product of critical factor ratings for each equipment. Any equipment that has a CFM value of 0 will be excluded from further analysis.

Step 2: determine the objective factors and compute objective factor measures. All objective factors must be expressed in terms of the same unit. Table 7.3 demonstrates step 2. The objective factor cost (OFC), C^{OF}, is the sum of all costs for an equipment; A is the sum of $(C^{OF})^{-1}$ for all equipment; the objective factor measure (OFM), M^{OF}, is given by the reciprocal of $C^{OF} \times A$. With this approach, the equipment with minimum cost corresponds to a maximum OFM value and the relationship of

EXAMPLE 7.1

ABC Plastics Inc. is a plastic products manufacturing company. The company has narrowed down the candidate material handling equipment to three different forklift truck models. The cost data for the three models is summarized below.

Cost category	*Model A*	*Model B*	*Model C*
Initial cost ($)	6000	7600	13 000
Economic life (years)	5	5	5
Annual maintenance or operating costs ($)	7800	7282	5720

Assuming that all models have a zero salvage value at the end of five years and an interest rate of 10%, select the best model.

Present worth method

In this method, all the costs are converted to present values and the total present costs of the candidates are compared. The candidate equipment with the least present cost is selected. The annual maintenance and operating costs (M&O) are converted to present values. This is done by multiplying the annual costs by the uniform series compound amount factor (P/A, 10%, 5). The value of this factor may be obtained from any standard book on engineering economics. All costs in terms of present worth are as follows.

Cost category	*Model A*	*Model B*	*Model C*
Initial cost ($)	6000	7600	13 000
Annual M&O cost ($)	29 568	27 605	21 683
Total ($)	35 568	35 205	34 683

Model C appears to be the best equipment based on the present worth method.

Annual cost method

In this method, all the costs are annualized and the total annualized costs of the candidates are compared. The candidate equipment with the least annual cost is selected. The initial investments are converted to annual costs. This is done by multiplying the annual costs by the capital recovery factor (A/P, 10%, 5). The value of this factor may be obtained from any standard book on engineering economics. All costs in terms of annualized costs are as follows.

Cost category	*Model A*	*Model B*	*Model C*
Initial cost ($)	1583	2005	3429
Annual M&O cost ($)	7800	7282	5720
Total ($)	9383	9287	9149

Model C appears to be the best equipment based on the annual cost method as well.

Internal rate of return method

Since all cash flows are costs, use of the internal rate of return (IRR) requires the application of incremental analysis. IRR is the interest rate at which the present worth of net cash flows is zero. The computation of the IRR is often an iterative procedure which may also require some interpolation. The difference in investment and M&O costs between models A and B and models B and C and the corresponding IRR values are as follows.

Incremental cost	*Model A to B*	*Model B to C*
Initial cost ($)	1600	5400
Annual M&O cost ($)	518	1564
IRR (%)	18.5	13.5

The sample calculation for model A to B is as follows:

$$-1600 + \$518\ (P/A, I\%, 5) = 0$$

This translates to $(P/A, I\%, 5) = 1600/518 = 3.1$. Looking through the uniform series compound amount factor tables, this value is for interest rates between 18% and 20%. Performing a linear interpolation yields an IRR of 18.5%. Model B is better than model A since the IRR of incremental investment is higher than the interest rate of 10%. Similarly, model C appears to be better than model B since the incremental IRR is higher than 10%. Hence, model C appears to be the best equipment based on the IRR method.

OFC for each equipment compared with other equipment is preserved in OFM.

Step 3: identify subjective factor measures and compute subjective factor weights. The following three steps are needed to compute subjective factor measures.

Step 3a: make a pairwise comparison of the subjective factors to determine subjective factor weights. If, for example, there are three subjective factors – say, ease of operation, safety and congestion – then there are three pairwise comparisons –

factor 1 with factor 2, factor 1 with factor 3, and factor 2 with factor 3. If factor 1 is preferred over factor 2, then assign 1 to factor 1 and assign 0 to factor 2. The results are illustrated in Table 7.4. The subjective factor weight (SFW) is computed as the total score for a factor divided by the sum of total scores for all factors.

Step 3b: make a pairwise comparison of the candidate equipment and compute the equipment weight (EW). This is done very similarly to step 3a but three times, once for each

Table 7.2 Systematic equipment selection procedure, step 1: an example

Critical factor	*Critical-factor rating for equipment*			
	1	*2*	*3*	*4*
Bidirectional movement	1	1	1	1
Weight capacity	1	1	1	1
Flow path (straight or flexible)	0	1	0	1
Useful life	1	1	1	1
Critical factor measure	0	1	0	1

Table 7.3 Systematic equipment selection procedure, step 2: an example

Objective factor	*Equipment*	
	2	4
Purchase cost ($)	1000	1100
Operating cost ($)	200	150
Maintenance cost ($)	50	50
Objective factor cost ($)	1250	1300
Objective factor measure	0.51	0.49

Table 7.4 Systematic equipment selection procedure, step 3a: an example

Composition	*Factor*		
	1	*2*	*3*
1 with 2	1	0	
1 with 3	1		0
2 with 3		1	0
Total score	2	1	0
Subjective factor weight	0.66	0.34	0

Table 7.5 Systematic equipment selection procedure, step 3c: an example

Subjective factor (SF)	*SF measure*	*Equipment weight*	
		equipment 2	*equipment 4*
Ease of operation	0.66	0.4	0.6
Safety	0.34	0.5	0.5
Congestion	0	0.7	0.3
Subjective factor measure		0.434	0.566

Table 7.6 Systematic equipment selection procedure, step 4: an example

Equipment	*Objective factor measure*	*Subjective factor measure*	*Final measure*
2	0.51	0.434	0.48
4	0.49	0.566	0.62

subjective factor. So, for the example with 2 equipment candidates and three subjective factors, there will be six equipment weights – two for each factor.

Step 3c: compute the subjective factor measure (SFM) by combining subjective factor weights and equipment weights as shown in Table 7.5. The total SFM is calculated as the sum of the product of the SFM and EW for each equipment.

Step 4: combine the OFM and SFM based on appropriate weights to obtain the final measure (FM). Assuming a subject factor weight of 0.4 and an objective factor weight of 0.6, the final measure for equipment 2 and 4 are given in Table 7.6.

Based on the values of FM, equipment 4 is selected as the final choice to meet the material handling requirements.

7.4 MATERIAL HANDLING EQUIPMENT ANALYSIS

In the preceding section, some approaches were presented for selecting material handling equipment. The approach assumed that candidate equipment types or models can be selected based on functional requirements. Analytical models are needed to verify if a particular equipment can meet the functional requirements. For instance, the loading time and the required part flow rate will to a great extent define the specifications – the length and speed of the conveyor and the spacing between the containers in the conveyor. Certain mathematical models are available to generate these specifications. In some cases the specifications are obtained from material handling equipment tables which have been developed from using empirical and experimental studies. In this section a few simple models for analysing conveyors and AGVs are presented.

7.4.1 Conveyor models

In this section, a quantitative analysis of the three different conveyor systems – single direction, continuous loop and recirculating – is presented (Groover, 1987).

Single direction conveyor systems

Let us consider a simple conveyor system with the loading and unloading stations separated by a distance of d m. Ignoring acceleration and deceleration effects, the conveyor operates at a speed of V m/min. Containers are loaded at one end and unloaded at the other. The travel time to transport a container from the load station to the unload station is d/V min. Let the loading time be L and the spacing between the containers be s. The time between the arrival of containers (s/V) must be greater than the loading time:

$$\frac{s}{V} \geq L \tag{7.2}$$

Equation (7.2) implies that the flow rate (V/s) is limited by the reciprocal of the loading time:

$$\frac{V}{s} \leq \frac{1}{L} \tag{7.3}$$

The time required to unload the conveyor must be less than or equal to the loading time. If the container can hold n parts then the part flow rate, R, on the conveyor system will be given as:

$$R = \frac{nV}{s} \leq \frac{n}{L} \tag{7.4}$$

This model assumes that there are no empty trips, the traffic factor is equal to 1 and that the conveyor does not stop for loading or unloading. An illustration is given in Example 7.2.

Loop conveyor systems

First, continuous loop conveyor systems are presented. This type of system has a forward loop and a return loop. The forward loop

EXAMPLE 7.2

A single direction conveyor system has to be designed to meet a part flow rate of 60 parts per minute. Each container in the conveyor can hold 6 parts. The distance between the loading and unloading points is about 10 m. Determine the required conveyor speed and the allowable loading time if the spacing between the containers is 1 m.

$$V = \frac{sR}{n} = \frac{1 \times 60}{6} \text{ m/min} = 10 \text{ m/min}.$$

$$\begin{aligned} L &\leq \frac{n}{R} \\ &\leq \frac{6}{60} \\ &\leq 0.1 \text{ min or } 6 \text{ s} \end{aligned}$$

carries parts from the loading station to the unloading station. The return trip simply brings the containers back to the loading point. Let us assume that the containers are spaced equally apart by a distance s and that each container can hold n parts. Let the forward loop distance be d and the return loop distance be e. Assume a conveyor speed of V m/min:

$$\text{time required to travel the conveyor loop} = \frac{d+e}{V}$$

$$\text{number of containers in the system} = \frac{d+e}{s}$$

$$\text{part flow rate, } R = \frac{nV}{s}$$

Since only the forward loop carries parts, the number of containers that carry parts is d/s and the number of parts in the system at any time is dn/s.

The second type of loop conveyor systems is known as a recirculating system since parts may not have to be unloaded at the unloading station and may be allowed to recirculate in the conveyor. In other words, the conveyor may be used for temporary storage. Some of the problems with recirculating conveyors are that no empty container may be available when needed at the loading point; similarly, no loaded container may be available at the unloading point when needed. Kwo (1958) provides a very good analysis of recirculating conveyors.

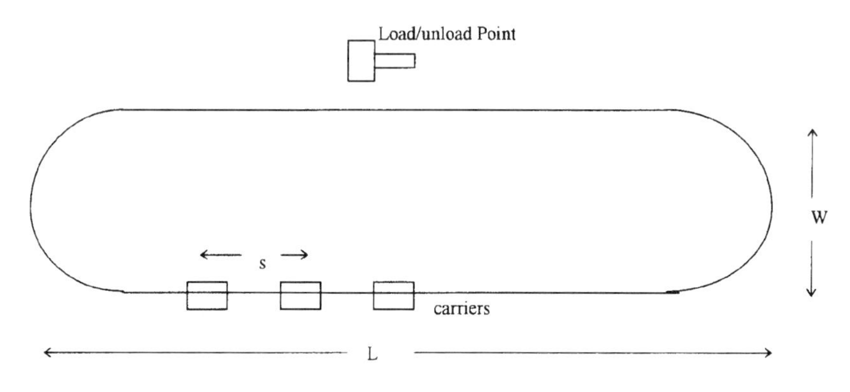

Figure 7.1 Carousel system: top view.

7.4.2 Carousel storage and handling systems

Carousel-type storage systems are very suitable for storing small parts and hence are very common among electronic and light assembly industries. Carousels may be floor mounted or may be hung from overhead rails. Rotation of the carousel may be controlled by using a hand lever, a foot pedal or a keyboard. A carousel consists of carriers that are spaced apart by a certain distance. Each carrier will have a certain number of bins. The location of a part in the storage system is identified by the carrier and bin number. Retrieval of a part to the load or unload point is done simply by entering the part number in the computer. The computer looks for its location (carrier and bin number) and rotates the carousel to bring the carrier or bin to the load or unload point. The top view of a carousel is shown in Figure 7.1.

In practice, an item may have to be retrieved from or placed in any random location. The average time to retrieve or place a part is computed as follows. Let the height of the carousel be H and the circumference of the carousel track, C, be given by:

$$C = 2L - 2W + \pi W$$

where L and W are the length and width, respectively. The number of carriers, n, is given by:

$$n = \frac{C}{s}$$

where s is the spacing between the carriers. The total number of bins, N, is given by:

$$N = nb$$

where b is the number of bins per carrier.

EXAMPLE 7.3

The oval of a top-driven carousel has a length of 40 ft and a width of 4 ft. The speed of the carousel is 10 ft/min. There are 60 carriers around the carousel and each carrier contains only one bin. For a single and bidirectional carousel, compute the average part retrieval time assuming a handling time of 30 s. Also, determine the spacing between the carriers.

$$C = (2 \times 40) - (2 \times 4) + 4\pi = 84.57 \text{ feet}$$

$$\text{Time for a single direction carousel} = \left(\frac{0.5 \times 84.57}{75} + 0.5\right) \text{min} = 1.064 \text{ min}$$

$$\text{Time for bidirectional carousel} = \left(\frac{0.25 \times 84.57}{75} + 0.5\right) \text{min} = 0.782 \text{ min}$$

$$\text{Spacing between carriers} = \frac{84.57}{60} \text{ feet} = 1.41 \text{ feet}$$

For a single-direction carousel of average distance L:

$$L = 0.5C + 0.5H$$

where $(0 + C)/2 = 0.5C$ is the time to get to the carrier, and $0.5H$ is the time to access the bin. Average time, T, is given by:

$$T = \frac{0.5C + 0.5H}{V} + M$$

where M is the handling time.

For bi-directional travel:

$$L = 0.25C + 0.5H$$

$$T = \frac{0.25C + 0.5H}{V} + M$$

An illustration is given in Example 7.3.

7.4.3 Automated guided vehicles

In this section, some of the design and operational problems encountered in installing and operating an AGV system are discussed. Designing the flow path layout and determining the capacity are the most important design problems. Flow path design involves making decisions about route segments to connect the machines, travel directions and the location of pick-up and drop-off points (Goetz and Egbelu, 1990). Capacity planning involves determining the number and types of AGVs to meet the material handling requirements (Kasilingam and Gobal, 1996). Vehicle routing and scheduling, load size specification and traffic management or control are the most important operational problems. Vehicle routing determines the optimal route between any two points based on the number of vehicles and the material handling requirements (Fujii and Sandoh, 1987). Vehicle scheduling or dispatching deals with the assignment of a vehicle to a task in real time (Egbelu and Tanchocho, 1984). The load size specification problem focuses on determining the optimal size of the unit load in order to minimize overall manufacturing costs (Steudel and Moon, 1987). Traffic management addresses zone release and control problems to minimize congestion and collision (Koff, 1987). In this section, a couple of methodologies to address the AGV capacity planning problem are presented.

AGV capacity planning

A couple of approaches to determine the number of AGVs needed to meet the material handling requirements will be presented in this section. The first approach is a simple analytical model proposed by Fitzgerald (1985). The second approach is a mathematical model that simultaneously determines the number of AGVs required and the assignment of AGVs to route segments (Kasilingam, 1991).

Analytical model

This model assumes constant vehicle velocity. The basic equation to determine the number of AGVs required, N^{AGV}, is as follows:

$$N^{AGV} = \frac{d}{C} \tag{7.5}$$

where d is the number of deliveries required per hour and C is the number of deliveries per hour per AGV. The number of deliveries required per hour is the expected material handling load on the system and is usually estimated based on manufacturing requirements. This information is an input to the model. The number of deliveries per hour per AGV is computed as follows:

$$C = \frac{60F}{T} \tag{7.6}$$

where F is the traffic factor and T is the total time per delivery per vehicle. The traffic factor represents the effect of traffic congestion, scheduling problems and other inefficiencies. The total time per delivery consists of loaded travel time, empty travel time and handling time. Handling time includes both loading time at the origin and

EXAMPLE 7.4

The proposed AGV system for a manufacturing unit must be capable of making 80 deliveries per hour. The type of AGV selected for this application has a velocity of 50 m/s. Based on the current facility layout, the average loaded travel distance will be 150 m and the average empty travel distance will be 100 m. The total loading and unloading time will average about 1 min. Assuming a traffic factor of 0.8, determine the number of AGVs required.

From equation (7.7), the total time per delivery per vehicle, T, is given by

$$T = \left(\frac{150}{50} + 1 + \frac{100}{50}\right) \text{min} = 6\,\text{min}$$

From equation (7.6), the number of deliveries per hour per vehicle, C, is given by

$$C = \frac{60 \times 0.8}{6} = 8$$

Equation (7.5) yields the number of AGVs required $= 80/8 = 10$

$$N^{\text{AGV}} = \frac{80}{8} = 10$$

unloading time at the destination. Travel times are a function of travel distance and AGV velocity:

$$T = \frac{L}{V} + H + \frac{D}{V} \tag{7.7}$$

where

L is the loaded travel distance (in meters);
D is the empty travel distance (meters);
V is the AGV velocity (meters per minute);
H is the handling time.

An illustration is provided in Example 7.4.

Mathematical model

The mathematical model is developed under the following assumptions:

- more than one vehicle type can service the material handling requirements between any two stations;
- the manufacturing system layout is known;

- the flow between stations and the loading and unloading times are known;
- the velocities, load carrying capacities and availability of different AGV types are known;
- all cost information is available.

The notation is as follows:

C_k is the annualized buying and maintenance costs for vehicle type k;
f_{jl} is the flow between stations j and l;
F_{jlk} is the cost of a loaded trip from j to l for vehicle type k;
C_{jlk} is the cost of an empty trip from j to l for vehicle type k;
G_k is the availability of vehicle type k in time units;
n_k is the capacity of AGV type k;
U_l is the unloading time at station l;
L_j is the loading time at station j;
t_{jlk} is the time for a loaded trip from j to l for vehicle type k;
r_{jlk} is the time for an empty trip from j to l for vehicle type k;
$N_k(j)$ is the net flow of vehicle type k at station j and is equal to $\sum_l (Y_{jlk} - Y_{ljk})$.

The decision variables are as follows:

X_k is the number of vehicles needed of type k;
Y_{jlk} is the number of loaded trips by vehicle type k from j to l;
W_{jlk} is the number of empty trips made by vehicle type k from j to l.

The objective function minimizes the sum of annualized, depreciated costs and loaded and empty vehicle travel costs. The first part on the right-hand side of equation (7.8) below represents the present worth of buying and maintenance costs. The second part is the loaded travel cost and the last part is the empty travel cost. This typically includes the cost of power, lubricants, maintenance labor and parts:

$$\text{minimize } z = \sum_k C_k X_k + \sum_j \sum_l \sum_k F_{jlk} Y_{jlk} + \sum_j \sum_l \sum_k C_{jlk} W_{jlk} \quad (7.8)$$

The model has four sets of constraints. The first set consists of demand constraints. These constraints ensure that the flow rate between any two stations is met by some combination of AGV assignments and are given as follows:

$$\sum_k Y_{jlk} \geq \frac{f_{jl}}{n_k} \quad \forall j, l$$

The second set of constraints are needed for flow balance. When $N_k(j)$ is greater than zero, empty trips of the kth vehicle will have to be made to station j from other stations. On the other hand, if $N_k(j)$ is less than zero then empty trips will have to be made from station j to other stations:

$$\sum_l W_{ljk} - \sum_l Y_{jlk} + \sum_l Y_{ljk} \geq 0, \forall j, k$$

$$\sum_l W_{jlk} + \sum_l Y_{jlk} - \sum_l Y_{ljk} \geq 0, \forall j, k$$

The third constraint set ensures that the available capacity is not exceeded. The total time of loaded and empty travel by vehicle k is limited by its availability and is represented as:

$$\sum_j \sum_l Y_{jlk}(U_l + L_l + t_{jlk}) - \sum_j \sum_l r_{jlk} W_{jlk} \leq G_k X_k, \forall k$$

The first part of the left-hand side of the constraint represents total loaded travel time (including loading and unloading) and the second part of the left-hand side of the constraint represents empty travel time.

The last constraint set includes non-negativity and integrality restrictions of the decision variables:

$$Y_{jlk}, W_{jlk} \text{ and } X_k \geq 0 \text{ and are integers}, \forall k$$

7.5 SUMMARY

Intrafacility logistics is most often treated separately and addressed as material handling along with manufacturing systems planning and design. Only recently has it been included as part of the overall logistics framework. The time and cost involved in moving or storing a product or component within a facility during its manufacturing or assembly is quite significant compared with its total manufacturing and assembly time and cost. The integration of intrafacility costs with other logistics costs will help to determine the best overall logistics supply chain. In this chapter, the various types of material handling equipment, methods to select equipment to meet the material handling requirements and some analytical models for planning material handling systems were presented. The models included conveyors, carousel systems and AGVs. The emphasis was more on AGVs because of their growing popularity among the high-technology industries.

PROBLEMS

7.1 Suggest the type of material handling equipment that is appropriate for the following situations:

automobile manufacturing;
road construction;
electronic assembly.

Explain the reasons for your suggestions.

7.2 What are the underlying assumptions of the various AGV capacity planning models presented in this chapter? Discuss the seriousness of the assumptions.

7.3 Discuss the interrelationship among material handling, building selection and layout decisions with suitable examples.

7.4 Discuss how each of the material handling principles outlined in this chapter can help to cut costs, increase productivity and help reduce accidents.

7.5 RGK Electronics is in the process of upgrading one of the assembly lines. The plan is to replace the old assembly bench with a new conveyor. The company has narrowed down its choices to two conveyor systems. The financial data for the two systems are given below:

	Conveyor A	*Conveyor B*
Purchase price ($)	40 000	50 000
Maintenance cost ($/year)	1000	800
Operating cost ($/year)	2000	1500
Life (years)	10	10

Assuming an interest rate of 12%, select the least cost system by means of the present worth method.

7.6 For problem 7.5, the following information is available on the two conveyors in terms of reliability, product damage and warranty. The relative weights for cost, reliability, damage and warranty are 0.4, 0.3, 0.2 and 0.1, respectively.

Factor	*Conveyor A*	*Conveyor B*
Reliability	0.8	0.95
Warranty (years)	5	4
Product damage (parts)	1 in 1000	1 in 1200

Use an appropriate method to select the best conveyor system.

7.7 Solve problem 7.5 by means of the internal rate of return method.

7.8 An overhead conveyor has hooks, and moves painted doors over a distance of 400 m in a single direction at 100 m/min between the paint station and the intermediate storage area. The cycle time for painting a door is 2 min. Determine the spacing between the hooks and the flow rate of parts in the conveyor system. Assume the unloading time is equal to the paint cycle time.

7.9 An automated guided vehicle system is used inside a telephone repair facility. The vehicles move repaired telephones from the repair facility to the packing area. Owing to the nature of the layout, the loaded travel distance is 300 m and the empty travel distance is 500 m. The system must be able to move 750 phones per hour. The load and unload time is about 1 min and the speed of the vehicle is 120 m/min. Determine the number of AGVs required.

7.10 What are the benefits of avoiding material handling? Can it be done? How? To what extent?

7.11 Develop a mathematical model to determine the requirements of AGVs over a period of five years. The model should consider the variations in part flow rates from period to period and changes in costs.

7.12 As a logistics engineer you have been assigned the responsibility of selecting the material handling equipment to move the finished pipes to the warehouse. Clearly describe your approach (including data requirements and procedure) to solve this problem.

REFERENCES

Egbelu, P.J. and Tanchocho, J.M.A. (1984) Characterization of automated guided vehicle dispatching rules in facilities with existing layouts. *International Journal of Production Research*, **22**, 359–74.

Fitzgerald, K.R. (1985) How to estimate the number of AGVs you need. *Modern Materials Handling*, **79**.

Fujii, S. and Sandoh, H. (1987) A routing algorithm for automated guided vehicles in FMS. *Proceedings of the 9th International Conference on Production Research*, Cincinnati, OH.

Goetz, W.G. Jr. and Egbelu, P.J. (1990) Guidepath design and location of load pickup/drop-off points for an automated guided vehicle system. *International Journal of Production Research*, **28**, 927–41.

Groover, M.P. (1987) *Automation, Production Systems, and Computer Integrated Manufacturing*, Prentice-Hall, Englewood Cliffs, NJ.

Kasilingam, R.G. (1991) Mathematical modeling of AGVS capacity requirements. *Engineering Costs and Production Economics*, **21**, 171–5.

Kasilingam, R.G. and Gobal, S.L. (1996) Vehicle requirements model for automated guided vehicle systems. *International Journal of Advanced Manufacturing Technology*, **12**, 276–9.

Koff, G.A. (1987) Automated guided vehicle systems: applications, controls, and planning. *Material Flow*, **4**, 3–16.

Kwo, T.T. (1958) A Theory of Conveyors. *Management Science*, 51–71.

Liang, M. and Dutta, S.P. (1988) An expert system for selecting material handling equipment types. Research report, Department of Industrial Engineering, University of Windsor, Windsor, Ontario.

Muther, R. and Haganas, K. (1969) *Systematic Handling Analysis*, Management and Industrial Research Publications, Kansas City, MO.

Riggs, J.L. (1982) *Engineering Economics*, McGraw-Hill, New York.

Steudel, H.J. and Moon, H.K. (1987) Selection of unit load size for automated guided vehicles in large cellular manufacturing environments. *Proceedings of the 9th International Conference on Production Research*, Cincinnati, OH.

Tompkins, J.A. and White, J.A. (1984) *Facilities Planning*, John Wiley, New York.

FURTHER READING

Apple, J.M. (1973) *Material Handling Systems Design*, The Ronald Press, New York.
Council of Logistics Management (1994) *Bibliography of Logistics Training Aids*, 2803 Butterfield Road, Suite 380, Oak Brook, IL 60521.
Gaskins, R.J. and Tanchoco, J.M.A. (1987) Flow path design for automated guided vehicle systems. *International Journal of Production Research*, **25**, 667–76.
Immer, J.R. (1953) *Material Handling*, McGraw-Hill, New York.
Muller, T. (1983) *Automated Guided Vehicles*, IFS, Bedford/Springer, Berlin.
Muth, E.J. (1972) Analysis of closed loop conveyor systems, *AIIE Transactions*, **4**, 134–43.

TRAINING AIDS: SOFTWARE AND VIDEOTAPES

Software

HOCUS Material Handling Modeling Software, P-E International, tel. (410) 239-3372.

Videotape

Greenbrier Intermodal, tel. (800) 826-0132: *AUTOSTACK Corporation Video.*
Penn State Audio-Visual Services, tel. (800) 826-0312: *Business Logistics Management Series: Materials Handling and Packaging in Logistics.*
Tompkins Associates, Inc., tel. (919) 876-3667: *Warehousing Strategies.*

CHAPTER 8

Transportation planning

8.1 INTRODUCTION

Transportation provides the link between production, storage and consumption. It also adds place value to a commodity. Availability of efficient transportation methods is the backbone of twentieth century logistics because of the increasing globalization of commerce and logistics. This is likely to continue through the next few centuries. The major requirements of the transportation industry to support global logistics are cost and transit time reduction, on-time delivery, lower variability of transit time, availability of seamless transportation service through a combination of modes, minimum delay, damage and loss and the availability of other options such as storage, pick-up and delivery and so on.

Figure 8.1 shows a schematic representation of the various parties involved in transporting a commodity from origin to destination or from shipper to consignee. The shipper has several options. He or she may:

use his or her own transportation equipment to send the commodity directly to the consignee;
send it through a common carrier;
send it through a forwarder.

A forwarder generally receives freight from several shippers, consolidates them and sends them using his or her own equipment or by using a common carrier.

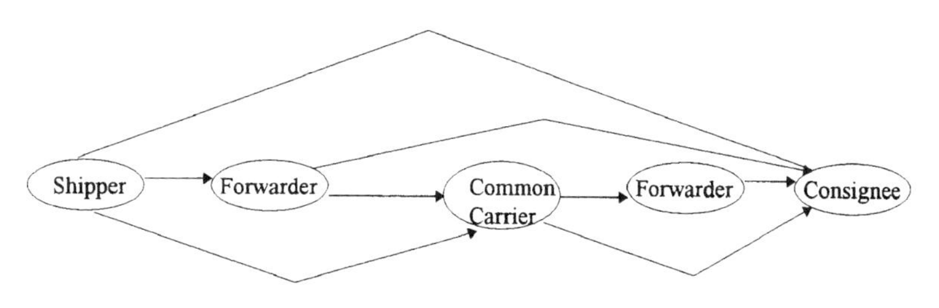

Figure 8.1 Transportation system structure.

In this chapter, a very brief introduction to different types of transportation systems and their components is first provided. Then, transportation costs and rates are discussed. Next, a series of models are presented which address some of the most common problems encountered in transportation planning such as mode selection, route selection, freight consolidation and vehicle routing and scheduling. Finally, four major transportation-related problems from the industry are discussed and the associated decision models and/or procedures are presented. The first model focuses on cargo revenue management from the airline industry which addresses the issue of managing rates and cargo space in order to maximize cargo profitability. The second model focuses on the sizing and control of the inventory of trailers in the retail industry in order to minimize the ratio of trailers to tractors. The third model focuses on the management of the distribution of locomotives from the railroad industry which attempts to minimize the total operating cost of locomotives but still power all the trains. The last model addresses one of the key problems that needs to be solved to develop a railroad operating plan: the blocking problem.

8.2 TRANSPORTATION SYSTEM

Facilities, equipment and people are the primary components of any transportation system. Facilities include terminals, tracks, bridges, tunnels, signals, roadways, waterways and docks. Equipment includes containers, cars, tractors, trailers, locomotives, aircraft and vessels. Loading and unloading crew, maintenance workforce, operating crew and other administrative and support staff come under the heading people resources. Railroads are financially responsible for building and maintaining their facilities. The railroads in the United States are owned and operated by many corporations. The eastern district is dominated by CSX, Conrail and Norfolk Southern. When approved by the Surface Transportation Board in late 1998 the eastern district will have only two key players – CSX and Norfolk Southern – both of which will acquire approximately half of Conrail. The western district is dominated by two major railroads – Burlington Northern–Sante Fe and Union Pacific–Southern Pacific.

Trucking companies build and maintain their terminals and docks but are not financially responsible for roads, tunnels and bridges. They do, however, pay for the roads, tunnels and other facilities through tolls and taxes. Water transportation may be classified into vessels and barges. Vessels are self-propelled and have a large holding capacity and travel very long distances, typically carrying freight from one country to another. Barges are generally small and are used for inland water transportation. Several barges are connected together (known as a 'tow') and are pulled by a tug. Air transportation facilities include signal stations, control towers and guide lights. Aircraft type may be narrow or wide body, the key difference being that only wide bodies can accommodate containers. Pipelines are popular forms of transportation to carry fluids. The facilities include pipes, pump stations and storage tanks.

Besides air, water, land and rail transportation systems another popular system is intermodal transportation. In intermodal transportation freight in a container or trailer moves along a combination of transportation modes from origin to destination. In most cases it involves ocean carriers, trucks and railroads. When air transportation is also involved freight may have to be transferred from one form of container to another. Intermodal transportation requires special types of material handling equipment to transfer containers between modes. In some cases, truck trailers are directly placed on flat rail cars. In other cases, containers are mounted on chassis while on road.

8.3 TRANSPORTATION ECONOMICS

As with any other enterprise the costs incurred in producing transportation services are associated with three elements: routes, terminals and vehicles. A route is the path over which the carrier operates and includes the right of way (land area being used) plus any roadbed and tracks or other physical facilities that are needed on the right of way. The nature of a route varies with the mode of transportation. In the case of railroads the route consists of the right of way plus roadbed, tracks, tunnels and bridges that are necessary. For highway carriers the routes consists of the right of way plus the highway or street. For water carriers the routes are lakes, oceans, rivers or canals. For air transportation the routes consist of the air corridor being used by the aircraft, which is often a part of the federal airway system.

Terminals are places where vehicles load and unload goods, make connections between points, make connections between routes within their own system and with other carriers and where vehicles are routed and dispatched. The size and complexity of terminal facilities varies with the mode of transportation. For example, water freight terminals or ports in domestic transportation are places where goods are transferred to and from vessels and where barges are assembled into tows. Vehicles of one kind or another are used in all modes of transportation in the analysis. Vehicles serve as carrying units and may be power units as well. The size and weight-carrying capacity of vehicles determine the size and weight of shipments. Unlike routes and terminals vehicles are usually not provided by the government but are provided by the operators.

These costs can be further broken down into many cost categories. For the purpose of this analysis the costs associated with providing transportation services shall be confined to only four types of costs: fixed, variable, attributable and non-attributable.

Fixed costs are costs that do not vary with output. They are unaffected by increases or decreases in the volume of traffic carried. They include various expenses for facilities, equipment, administration, interest on investment, insurance and taxes. They are called fixed costs because they are incurred regardless of whether any traffic is carried. For example, locks and dams must be maintained and manned without regard to volume of river traffic.

In contrast to fixed costs, variable costs fluctuate directly with the magnitude of traffic volume. Variable costs are incurred solely to produce transportation services needed to carry traffic. They include not only directly measurable costs but also those arising from wear and tear on the equipment used in performing transportation service. Examples of variable costs include direct labor cost, fuel costs and operating supplies. When a system has high fixed costs there is a good opportunity for economies of scale. Railroads, barge lines and pipelines have a very high fixed cost whereas trucks and airlines have a very high variable cost.

Attributable and non-attributable costs are used mainly in cost responsibility determination related to highway transportation. Attributable cost items are related to vehicle characteristics but not to functions of travel distances. Examples of attributable costs include bridge construction costs, rehabilitation and replacement costs of bridge structural elements, culvert construction costs, highway sign structure costs, pavement construction costs, pavement rehabilitation costs and so on.

Non-attributable costs are also known as common costs. They refer to expenditures which result from non-traffic-related causes such as the action of environmental forces (including weather, aging effects, salt and other chemical agents) and expenditures that are incurred based upon safety or aesthetic considerations. These costs cannot be attributed specifically to any particular user class or group of user classes. Examples of non-attributable cost items for highways are snow removal, street lighting and cleaning, traffic signals, drainage maintenance, guardrails and so on.

8.3.1 Transportation cost modeling

The pricing of a transportation service is done very similarly to the pricing of any manufactured or assembled product. Total transportation cost is the sum of fixed, variable and overhead costs. A certain profit margin is added to the total cost to determine the transportation rate or price for the service. As discussed before, facilities and equipment costs fall under fixed costs. Maintenance and operating costs fall under variable costs. Variable costs depend upon line haul charges and handling costs at the origin and destination. Line haul charges vary based on distance and weight, and handling costs depend upon weight.

Transportation providers consider only the fixed and variable costs associated with providing the service and charge their customers based only on these costs. However, there are other cost elements that need to be considered if one tries to estimate the true cost of providing the service. In this section, a model is presented which addresses the costs of providing a transportation service (Hay, 1989).

General transportation cost model

The model is specified as follows:

$$C_{tot}^{trans} = C^{cap} + C^{op} \tag{8.1}$$

where

C_{tot}^{trans} is the total transportation cost;
C^{cap} is the capital costs;
C^{op} is the operating costs.

Capital costs and operating costs are defined as:

$$C^{cap} = C^{fac} + C^{equip} \tag{8.2}$$

$$C^{op} = C^{FM} + C^{EM} + C^{trans} + C^{traf} + C^{gen} \tag{8.3}$$

where

C^{fac} is the facilities cost;
C^{equip} is the equipment cost;
C^{FM} is the facilities maintenance cost;
C^{EM} is the equipment maintenance cost;
C^{trans} is the transport cost;
C^{traf} is the traffic cost;
C^{gen} is the general cost.

These cost components are as follows:

- capital cost – the costs of providing initial plant and equipment and additions to or betterment of those facilities, such as initial roads construction, railroad tracks construction, ports construction, etc; this includes interest charges on the capital invested;
- operating cost – the remaining costs of providing transportation services;
- facilities cost – investment in route, structure and terminals;
- equipment cost – investment in vehicles;

- facilities maintenance cost – the cost of maintaining roadway and track, pavement and subgrade, rivers and harbors, channels and dams, pipelines, and so on.
- equipment maintenance cost – all costs of maintaining motive power and rolling stock such as cars, locomotive, trucks, airplanes, etc.
- transport cost – the costs of conducting transportation, such as power and fuel, the wages of the vehicle crew and the wages of those directing vehicle movements;
- traffic cost – the costs of traffic solicitation, the wages of highway safety officers, advertising, publishing rates and tariffs and administration;
- general cost – the costs of general office expenses, legal advice, accounting and the salaries of general officers and their staffs.

Pricing internal transportation services

Several large companies have separate transportation departments and also own their transportation fleet. All related costs of owning and operating the fleet for providing the necessary services come from the transportation department budget. Hence it becomes very important to price appropriately the transportation service provided to the users within the company. Three major methodologies can be used for pricing an internal transportation service. These are: cost-based, market-based and a combination of the two.

Cost-based transfer pricing

Cost-based pricing can be done in three different ways. One approach is to charge the actual full cost to the internal user. In this case, the user pays for all actual costs, fixed and overhead. Also, since actual costs are charged, operating inefficiencies are passed on to the user. Another approach is to charge only the standard full cost, in which case operating inefficiencies are not passed on to the users. The third approach is known as the marginal or prime cost method. In this case, fixed costs are treated as overheads and only variable costs are charged. This approach is useful when excess capacity exists. It should also be noted that fixed costs may increase beyond a certain volume.

Market-based transfer pricing

This methodology can be implemented in two ways. The first approach is to charge what a competitive outside carrier would charge for a similar service. This approach should consider the differences in service and rates. The market price may be higher or lower than the actual cost. Excess capacity in the market place may lower price. Under this approach market price must be monitored constantly. The second approach is to charge an adjusted market price (market price – cost savings). The adjusted market price may be lower if the in-house transportation department is more efficient and may be higher otherwise.

Transfer pricing based on a combination

This can be achieved in two ways. The first method is based on a negotiated price agreed upon between the transportation department and the users. For the negotiation to be more effective a comparable market price must exist and the user must have the flexibility to use outside carriers. The second approach is based on the target profit of the transportation department. Under this method, price will be equal to the actual or standard cost plus the target profit for the department.

Computing trip costs

Computing the cost of a trip is very fundamental in transportation pricing. Trip cost is computed by adding three cost components, each of which is dependent upon different attributes. The first component is the load-based cost, the second is the time cost component and the last is the distance-based component.

$$\begin{aligned}\text{cost per trip} = {} & (\text{costs per load}) + (\text{functional costs per hour}) \\ & \times (\text{total trip hours}) + (\text{functional costs per mile}) \\ & \times (\text{total trip miles})\end{aligned} \tag{8.4}$$

The cost per load is estimated by using historical overhead costs and loads. The functional cost per hour is computed by adding driver wages, interest, depreciation and lease expenses and facility expenses and dividing the sum by the total hours spent by crew and equipment usage. The functional cost per mile is calculated by adding the costs of fuel and oil, equipment maintenance, tires and tubes and of insurance and accident cover and dividing the sum by the number of loaded and empty miles.

Special costing considerations

Assignment of empty miles to a trip and adjusting for cube or density factors are two most important special transportation costing issues. Assignment of empty miles to a trip may be done in any of three ways as follows. One approach may be to add post-deadhead miles (empty miles incurred after the loaded trip) to the loaded travel distance; another is to add the pre-deadhead miles; and the third is to add 50% of pre-deadhead miles and 50% of post-deadhead miles. Mixed loads in a trip pose problems in allocating costs to different loads. One example is when part of the load is computer sheets and the other part is packing material. The following method is generally used for this situation. First, compute standard density by dividing the truck payload by trailer volume. Then, convert product volumes to weights by using the standard density. Finally, use the maximum of {weight based on standard density, actual weight}.

8.3.2 Transportation rates

Transportation rates depend upon commodity type, weight, distance, level of service and other options requested. The rate for a bulk commodity may be less than the rate for a fragile commodity. Low-density commodities may have higher rates per hundredweight than high-density commodities. The level of service desired or requested by the shipper may add some premium to the rates. For instance, three-day delivery may have a higher rate than a five-day delivery. Similarly, same-day shipping may cost more than next-day shipping. Options requested may include pick-up and/or delivery, partial delivery or multiple pick-ups for a single shipment. Different industries and different companies within the same industry may use different types of rate profiles. The type of profile used may depend upon the type of services offered, the distribution of cost components (relationship between handling versus transportation) and the type of pricing policies used in order to meet certain strategic objectives such as capturing market share as opposed to profitability.

Weight-based rates

These rates vary with the weight of the freight shipped and not with distance. Examples include postal rates and some of the courier and overnight service rates. This type of rates is simple and easy to use. Most of the costs in this type of service are handling related. The rate changes occur at certain weight break points. For instance, the first 4 oz will be charged $7 for overnight delivery within the United States. In most cases there is a minimum charge; when the total freight charge based on weight rate is less than the minimum charge the minimum charge is levied.

Distance-related rates

In this case the rates vary with distance and weight. For a given weight, the rates vary with distance, either directly in a linear fashion or in a non-linear manner. Most of the line haul rates vary directly with distance since the key expenses are related to fuel and labor; fuel varies with distance and labor varies with time, which depends upon distance. When handling costs are included, transportation rates increase at decreasing rates since handling costs are spread over the entire distance.

EXAMPLE 8.1

A particular product is manufactured by two companies. The selling price for the product is $1 per unit. Suppose the production cost of the manufacturer located in A is $0.75 per unit and the transportation cost from A is $0.20 per unit. If the production cost of the manufacturer located in B is $0.85 per unit then the maximum this manufacturer will be willing to pay for transportation is only $0.15 per unit.

EXAMPLE 8.2

Suppose the current rates for shipping a 6 oz packet from Dallas, TX, to New York is $6 and from Dallas to Raleigh, NC (Raleigh is in between Dallas and New York) is $5. A new player enters the small-packet market and offers a rate of $4.50 to ship a 6 oz packet from Dallas to New York. The current players will then be forced to offer a rate of $4.50 to New York. They will also be forced to blanket the rate to Raleigh to $4.50 to be consistent with the new New York rates.

Demand-related rates

These rates depend neither on weight nor on distance. They are dictated by external market conditions.

Examples 8.1 and 8.2 serve as illustrations.

Contract rates

These are specially agreed upon rates between a shipper and a carrier. They are usually based on the promised volume and duration of business, service reliability, historical data on actual business provided, type of commodity, traffic corridor involved and image of the shipper.

Line haul rates

Line haul rates vary with distance, weight and commodity type. They do not include prices for optional services offered such as pick-up, drop-off, storage, etc. Commodities are grouped into classes based on the value, liability, hazard potential and handling requirements of the commodity. For instance, according to the Uniform Freight

Classification, window shade cloth or felt in boxes are rated as class 100 whereas ramie cloth in bales or rolls are classified as class 55. Line haul rate tables contain the rate per hundredweight for various combinations of distance and commodity classes. For example, the first row in the table may contain rates for all classes up to 40 miles. The first column in the table may contain rates for class 400 for various distance ranges. To compute the rate per hundredweight, one must first identify the class based on the Uniform Freight Classification then look at the cell entry in the rate table corresponding to the distance and class. If the rate for a given distance is known for class 100 the corresponding rate for any other class may be determined as follows:

$$\text{rate for class 150} = \text{rate for class 100} \times \frac{150}{100}$$

Similarly,

$$\text{rate for class 55} = \text{rate for class 55} \times \frac{55}{100}$$

Other special rates

Special rates may cover a certain period, region or commodity. They may be higher or lower than the normal rates. Rates during certain off-peak periods, from or to certain origins or destinations or for some special commodities are examples of these. Larger shipments (truckload or carload) may have lower rates than less-than-truckload or less-than-carload. Light and bulky shipments may have higher rates than normal or dense shipments. Shipments originating and/or terminating in a foreign country may have a better rate than an entirely domestic shipment. Shipments with a lower service level (next-day delivery) and liability coverage may have lower rates than shipments with a higher service level (same-day) and coverage. Special or additional services may also increase the regular transportation rate; these services may include changing the destination while enroute, changing the consignee after the destination is reached, in-transit storage and partial loading or unloading at intermediate points.

8.4 GENERAL TRANSPORTATION MODELS

In this section, an overview is provided of some of the basic transportation models. A transportation system may be represented in the form of a network of nodes and arcs; the nodes typically represent cities, airports, stops or depots. The arcs represent the links or routes between the nodes. Nodes and arcs may have capacity limitations. Five most commonly encountered transportation problems are mode selection, carrier routing, fleet sizing, scheduling and shipment consolidation.

8.4.1 Mode selection

Alternative modes are usually available to move shipments between any two points. The mode selection problem focuses on selecting the best mode to transport a shipment considering mode capacities, costs, handling requirements and distances. Several factors are important in selecting the transportation mode – frequency of service, speed, transit time, transit time variability, cost, availability, safety, security and customer service. Factor analysis (section 4.3.1), weighted factor analysis (section 4.3.2), and analytic hierarchic process (section 4.3.4) can also be used for selecting mode. The following examples illustrate some of the methods available for mode selection. These methods assume that a single mode is used to transport a shipment from origin to destination. The problem becomes more complex when more than one mode is involved and intermodal transfer costs are significant. The complexity is further increased when transfer costs vary depending upon the transfer point and are based on incoming and outgoing mode at a transfer point. Intermodal transportation planning is discussed in section 8.4.6. An illustration of mode selection is provided in Example 8.3

EXAMPLE 8.3

A mineral processing plant has short-listed two transportation modes as potential candidates for transporting the ore from the mines to the plant. The evaluation of the two modes with respect to a few critical factors is given below (higher scores indicate better site).

Factor (weight)	*Mode 1*	*Mode 2*
Transit time (8)	7	5
Transit time variability (7)	5	7
Cost (3)	8	4
Capacity (6)	2	6
Loss or damage (7)	5	4
Customer service (5)	6	9

Determine the best mode for transporting the ore from the mines to plant.

The weighted rating for mode 1 is 192, and the weighted rating of mode 2 is 210. Since higher scores imply better mode, the ore must be transported using mode 1.

Since transit time depends upon the type of transportation mode, mode selection decisions also influence in-transit inventories, inventories at the source (plant) and at the consumption centers. A faster transportation mode requires less inventory. This is illustrated in Example 8.4.

8.4.2 Route selection

Alternative routes are generally available to go from one place to another. Carrier routing focuses on determining the shortest path to transport a shipment from origin to destination and may be different for different modes. The shortest path may be in terms of distance, time or cost. Carrier routing generally follows mode selection; however, it is best to solve the mode and carrier routing problems jointly since the availability of routes may depend upon the selected mode. Vehicle routing problems may be classified into the following categories:

- intermediate cities along the route and origin and destination are different;
- intermediate points are different but origin and destination are the same;
- multiple origins and multiple destinations with no intermediate points;
- multiple origins and multiple destinations with intermediate or transshipment points.

Origin and destination are different

Common applications of this include routing a shipment from origin to destination through a transportation network without considering other aspects such as route capacity and side constraints. This problem is commonly solved by well-known shortest path algorithms. A label-based shortest path algorithm consists of the following four steps.

Step 1: identify solved and unsolved nodes in the network;
Step 2: for each solved node, identify the directly connected nodes that are unsolved;
Step 3: select the unsolved node with smallest distance from the solved node; include that unsolved node in the list of solved nodes;
Step 4: stop if destination is reached; else, go to step 2.

Example 8.5 illustrates the use of a label-based shortest path algorithm.

Origin and destination are the same

This problem considers routing a vehicle from a depot to serve certain customer locations and return to the depot. Common applications

EXAMPLE 8.4

Wall-Hall paper company manufactures a certain grade of paper used for news printing. Bales of paper are shipped from the plant to a regional warehouse close to customer locations. Currently, goods are transported by rail. The average transit time from plant to regional warehouse is about 10 days. To ensure good customer service the company has to maintain an average inventory of 10 000 bales at the warehouse. The company is interested in knowing the impact of using trucks on total costs. Truck transportation will bring down the transit time by 3 days; every day saved in transit time results in 2% reduced inventory. The following additional information is available:

cost of rail transportation \$0.2 per bale;
number of shipments to meet the demand = 10;
cost of truck transportation = \$0.3 per bale;
number of shipments to meet the demand = 20;
cost of carrying inventory = \$6 per bale per year;
annual demand = 100 000 bales.

Determine if it is worthwhile to switch to truck transportation.

The annual cost of rail transportation is as follows:

$$\text{transportation cost} = (\$0.2 \text{ per bale}) \times (100\,000 \text{ bales}) = \$20\,000$$

$$\begin{aligned}\text{warehouse inventory cost} &= (\$6 \text{ per bale per year}) \times (10\,000 \text{ bales}) \\ &= \$60\,000\end{aligned}$$

$$\begin{aligned}\text{in-transit inventory cost} &= (\$6 \text{ per bale per year}) \times (100\,000 \text{ bales}) \times \left(\frac{10}{365}\right) \\ &= \$16\,438\end{aligned}$$

$$\text{total cost} = \$96\,438$$

The annual cost of truck transportation is as follows:

$$\text{transportation cost} = (\$0.4 \text{ per bale}) \times (100\,000 \text{ bales}) = \$40\,000$$

$$\begin{aligned}\text{warehouse inventory cost} &= (\$6 \text{ per bale per year}) \times (5000 \text{ bales}) \times (0.94) \\ &= \$28\,200\end{aligned}$$

$$\begin{aligned}\text{in-transit inventory cost} &= (\$6 \text{ per bale per year}) \times (100\,000 \text{ bales}) \times \left(\frac{7}{365}\right) \\ &= \$11\,506\end{aligned}$$

$$\text{total cost} = \$79\,706$$

The total cost analysis shows that switching to trucking will result in an annual cost saving of \$16 732.

EXAMPLE 8.5

ABC Trucking Company is in the business of transporting furniture from the factories in city A to the wholesale stores in city B. The transportation network consisting of the intermediate cities is as follows:

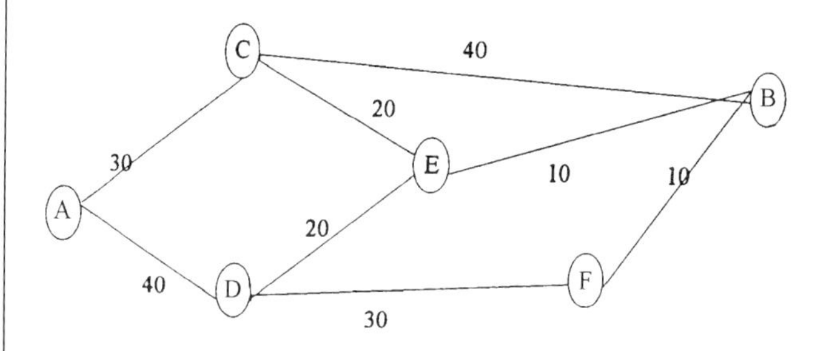

The values along the links show the distances between adjacent cities. Determine the shortest path from A to B.

The four-step algorithm described above is iteratively applied to the network shown above and the results are as follows:

Iteration	*Solved nodes*	*Directly connected unsolved nodes*	*Total distance*	*Shortest distance and connection*
1	A	C	30	30, AC
		D	40	
2	A	D	40	40, AD
	C	B	$30 + 40 = 70$	
		E	$30 + 20 = 50$	
3	C	E	$30 + 20 = 50$	50, CE
	C	B	$30 + 40 = 70$	
	D	E	$40 + 20 = 60$	
	D	F	$40 + 30 = 70$	
4	C	B	$30 + 40 = 70$	60, EB
	D	E	$40 + 20 = 60$	
	D	F	$40 + 30 = 70$	
	E	B	$50 + 10 = 60$	

From the computations shown in the Table, it can be seen that the shortest path to go from A to B is through C and E.

include local delivery trucks for bottled drinks, parcels and milk to stores within a region. This type of problem falls under the umbrella of the 'traveling salesperson problem' (TSP) in operations research. The objective is to determine the sequence of customer locations to visit

EXAMPLE 8.6

A milk delivery truck from depot 4 has been assigned to serve three customer locations. Distances in miles between the depot and the customer locations are as follows:

From	*To*			
	4	*B*	*C*	*D*
4		22	31	45
B	22		18	27
C	31	18		38
D	45	27	38	

Determine the best routing to serve the customers.

Step 1: customer location B is closest to the depot;
Step 2: customer location C is closest to customer location B;
Step 3: all locations have not been selected;
Step 2: the remaining location D is after location C along the route before returning to the depot.

The total distance for this routing (depot–B–C–D–depot) is 123 miles.

before returning to the depot in order to minimize the total distance traveled. Here, a simple greedy approach to solve this problem is presented.

Step 1: select the customer location that is closest to the depot;
Step 2: from the locations that have not been selected yet, select the next location that is closest to the most current selected location;
Step 3: if all locations have been selected, stop; else, go to step 2.

An illustration is provided in Example 8.6.

Multiple origins and destinations with no intermediate points

This problem is concerned with optimally distributing the available supply from multiple sources to the demand at various customer locations. Common applications of this include distribution of manufactured goods from plants to warehouses, supplying customers from warehouses, etc. This type of problem is known as the 'transportation problem' in the literature. Formulation of the transportation model and its solution using Vogel's method are illustrated in section 5.4.2.

EXAMPLE 8.7

Calicut Supplies Inc. manufactures valves at two plants in the east coast of the United States, one in city A and one in city B. The plant at city A has a production capacity of 150 valves per day, and the plant at city B has a capacity of 200 valves per day. The valves are shipped by truck to the demand points, cities C and D, on the west coast. The daily demand for valves is about 130 both at city C and at city D. The company also operates two intermediate trans-shipment points at cities E and F to consolidate shipments when it is appropriate to do so. The unit costs of shipping valves are as follows:

From city:	*To city:*					
	A	*B*	*E*	*F*	*C*	*D*
A	0	13	4	6	12	14
B	13	0	7	6	13	12
E	4	7	0	3	8	8
F	6	6	3	0	7	8
C	12	13	8	7	0	17
D	14	12	8	8	17	0

Determine the best routing for shipping valves from plants to demand points.

The problem can be solved in two phases. The first phase is to convert the trans-shipment problem into a transportation model by means of the following steps.

Step 1: add a dummy row or column to balance the problem in terms of supply and demand. Since total supply is 350 valves and total demand is only 260 valves, a dummy column with a demand of 90 valves is added;

Step 2: construct a transportation table by including all the cities (origin, destination and intermediate) as supply and demand points. This will result in a 6×7 table including the dummy column;

Step 3: this step helps to determine the demand and supply for all the points by using the following set of rules:

Nature of point in trans-shipment problem	*Value of supply in transportation table*	*Value of demand in transportation table*
Supply point	Original supply + total supply	Total supply
Trans-shipment point	Total supply	Total supply
Demand point	Total supply	Original demand + total supply
Dummy point	0	Original supply – original demand

The total supply in this case is 350 valves. The resulting transportation table is as follows:

From city:	*To city:*							
	A	*B*	*E*	*F*	*C*	*D*	*Dummy*	*Supply*
A	0	13	4	6	12	14	0	500
B	13	0	7	6	13	12	0	550
E	4	7	0	3	8	8	0	350
F	6	6	3	0	7	8	0	350
C	12	13	8	7	0	17	0	350
D	14	12	8	8	17	0	0	350
Demand	350	350	350	350	480	480	90	

The second phase involves solving the transportation model by any of the known methods, such as Vogel's approximation procedure (section 5.4.2).

Multiple origins and destinations with intermediate points

This problem is concerned with optimally distributing the available supply from multiple sources to the demand at various customer locations with an added flexibility to consolidate or distribute at intermediate locations. Some of the origins and destinations may also serve as intermediate locations. This problem is known as the 'trans-shipment model' in the literature. The first step in solving a trans-shipment problem is to formulate it as a transportation model with use of a few rules. Once formulated it can be solved by using any of the available transportation model algorithms. Example 8.7 illustrates the application of the trans-shipment model.

8.4.3 Fleet sizing

Fleet sizing decisions determine the number and type of vehicles needed to meet the transportation demand at the lowest cost but still deliver the expected level of customer service. As the size of fleet increases customer service in terms of response time and frequency of delivery improves; however, vehicle utilization decreases. Here, some very simple models are presented for addressing the following two cases of fleet sizing: a homogeneous fleet (a fleet with only one type of vehicle) and a non-homogeneous fleet.

Homogeneous fleet sizing model (Wyatt, 1961)

The notation used is as follows:

F is the fixed cost per day (this is incurred regardless of vehicle usage and it typically includes depreciation, road tax and basic driver wages);
V is the variable cost per day (this is incurred only when the vehicle is in use, and includes fuel, tires, etc.);
H is the hiring cost per vehicle per day;
Y is the number of working days in a year;
N is the number of vehicles in the fleet;
P is the number of extra days of work made possible by the additional vehicle.

The model is specified as follows:

$$\text{hiring cost} = PH$$

$$\text{incremental vehicle cost} = FY + PV$$

It makes sense to buy an extra vehicle when

$$FY + PV < PH$$

or, alternatively, when

$$\frac{F}{H - V} < \frac{P}{Y}$$

Non-homogeneous fleet sizing model

This model is a simple extension of the linear programming formulation of the homogeneous fleet sizing model proposed by Wyatt (1961). The notation is as follows:

X_{jk} is the number of company vehicles of type k required on day j;
h_{jk} is the number of hired vehicles of type k required on day j;
d_{jk} is the demand for loads equivalent to vehicle type k on day j;
F_k is the fixed cost of vehicle type k per day;
V_k is the variable cost of vehicle type k per day;
H_k is the hiring cost of vehicle type k per day.

The model is specified as follows:

$$\text{minimize } z = \sum_j \left[\sum_k X_{jk}(F_k + V_k) + H_k \sum_k h_{jk} \right] \tag{8.5}$$

subject to

$$X_{jk} + h_{jk} \geq d_{jk}, \forall\, j, k$$

$$X_{jk}, h_{jk} \geq 0 \text{ and are integers}, \forall\, j, k$$

The objective function in equation (8.5) minimizes the fixed and variable cost of owning company vehicles and the variable cost of hiring additional vehicles. The first constraint ensures that each day all loads demanded are carried by the appropriate vehicle type. The second constraint represents the integrality of the decision variables.

Fleet assignment model

A closely related model and a variation of the fleet sizing problem is the fleet assignment problem. In this section a simple fleet assignment problem is given as an assignment model. The notation is as follows:

$$X_{jk} = \begin{cases} 1 & \text{if truck type } j \text{ is assigned to trip } k, \\ 0 & \text{otherwise;} \end{cases}$$

c_{jk} is the cost of assigning truck type j to trip k (this will be a very high value if a truck type cannot be assigned to a trip).

The model is as follows:

$$\text{minimize } z = \sum_j \sum_k X_{jk} c_{jk} \tag{8.6}$$

subject to

$$\sum_k X_{jk} = 1, \forall j$$

$$\sum_j X_{jk} = 1, \forall k$$

and X_{jk} is a binary integer, $\forall j, k$.

The objective function, represented in equation (8.6), minimizes the total cost of assigning vehicle types to trips. The first constraint ensures that each truck type is assigned to one trip. The second constraint ensures that each trip gets a truck. The last constraint ensures the binary nature of the decision variables.

The fleet assignment model presented above can be solved by using the Hungarian method (Kuhn, 1955). Example 8.8 illustrates the use of assignment modeling and the Hungarian method.

8.4.4 Vehicle scheduling

Whereas routing decisions involve the sequence of stops to go from origin to destination based on distance, scheduling decisions consider a lot more issues. The primary issue is time, an additional element of importance in transportation – such decisions focus on how many vehicles should be dispatched from where to where on what day and at what times. Other issues are constraints on fleet size, available routings based on type of equipment or vehicle, volume to be picked up or delivered at each city, time windows for service (restrictions on when a

EXAMPLE 8.8

A company has two 8 ton trucks and two 9.5 ton trucks. There are four trips that need to be made in a day. The cost of assigning the trucks to trips are as follows:

Truck (weight)	*Trip*			
	1	*2*	*3*	*4*
1 (8 ton)	14	5	8	7
2 (8 ton)	2	12	6	5
3 (9.5 ton)	7	8	3	9
4 (9.5 ton)	2	4	6	10

Determine the assignment of trucks to trips that will minimize the overall cost.

The problem is solved by the Hungarian method.

Step 1: find the minimum cost cell for each row. Construct a new cost matrix with the new cell cost equal to the cell cost minus the minimum cost for that row. Now, for the new matrix, find the minimum cost for each column. Construct a new matrix with the new cell cost equal to the cell cost minus the minimum cost for that column:

Truck	*Trip*			
	1	*2*	*3*	*4*
Cost matrix after subtracting row minimum:				
1	9	0	3	2
2	0	10	4	3
3	4	5	0	6
4	0	2	4	8
Cost matrix after subtracting column minimum:				
1	9	0	3	0
2	0	10	4	1
3	4	5	0	4
4	0	2	4	6

Step 2: draw the least number of lines (horizontal and/or vertical) to cover all the zeros in the matrix. If the number of lines equals the number of rows or columns, stop and obtain the optimal assignment from the matrix by picking the cells with zero costs. Otherwise, go to step 3. This step is shown above for the second cost matrix above obtained after subtracting the column minimum. From this matrix it can be seen that the number of lines (3) is less than the number of rows or columns (4);

Step 3: find the smallest non-zero value not covered by the lines, a. Subtract a from each element uncovered by the lines and add a to each element covered by two lines. Return to step 2. For this example $a = 1$.

Truck	*Trip*			
	1	*2*	*3*	*4*
1	10	0	3	0
2	0	9	3	0
3	5	5	0	4
4	0	1	3	5

Step 2: the number of lines is equal to four. Hence, we stop and make the assignment based on zero cost cells. The optimal assignment is truck 1 to trip 2, truck 2 to trip 4, truck 3 to trip 3 and truck 4 to trip 1. In other words, type 1 trucks are assigned to trips 2 and 4 and type 2 trucks to trips 1 and 3.

customer may be served) and crew rest and work periods based on safety considerations.

Basic principles for good scheduling (Ballou, 1992)

The basic principles are to:

- assign a vehicle to stops or cities that are close to each other, taking vehicle capacity into consideration (spatial coordination);
- combine deliveries and pick-ups at a stop whenever possible;
- combine deliveries on same day of the week together (temporal coordination);
- build routes beginning with the farthest stop from the depot and to include cities working backwards;
- use the largest vehicle available first to minimize the cost of underutilization;
- avoid narrow time windows; renegotiate them with customers;
- explore alternate means of delivery or pick-up to or from remote or low-volume locations.

Solution procedures for vehicle scheduling

The solution strategies for vehicle scheduling include the consideration of cluster first, route second; route first, cluster second; savings and insertion; improvement or exchange; and mathematical models. In the cluster first, route second approach the demand nodes are first grouped into clusters. Each cluster contains a set of nodes that are close to each other and satisfy constraints on vehicle capacity and time windows. Then, the best routing is determined for each cluster. The route first, cluster second method works in the reverse sequence. First a large single route is constructed to include all demand nodes. The route is then partitioned into a number of smaller, but feasible, routes. In the savings and insertion approach two nodes are combined into one cluster by comparing the savings with other available alternatives. While combining nodes one should check to ensure that the total travel time or distance does not violate constraints on maximum crew time or some other time sensitive service and that the combined load does not exceed the vehicle capacity. Also, the selected city or stop should be from the set of cities or stops not already combined. In the improvement method, at every step a feasible solution is altered to yield another feasible solution with a lower cost. The procedure continues until no additional cost reduction is possible. Mathematical models may be linear or non-linear, depending upon the constraints. Most models are based on a time–space network – each node in the network represents a place and point in time. Solution procedures to mathematical models depend upon the size and complexity of the formulation – accordingly, optimal or near-optimal heuristic procedures are employed.

8.4.5 Shipment consolidation

Shipment consolidation may be done by time, inventory, and space or by equipment. All of these approaches attempt to minimize the operating expenses by cutting down on the number of trips. Shipment consolidation may lead to lower service level, higher inventory costs and additional handling costs. This is usually more than offset by the reduced equipment, crew and maintenance costs.

Time-based consolidation

When two customers at the same location or at nearby locations want their deliveries on different days, say one on Monday and the other on Tuesday, there are probably three options available. The first option is to deliver to the first customer on Monday and to the second on Tuesday, resulting in two trips but just-in-time deliveries. The second option is to deliver to both of them on Monday, resulting in excess inventory at the second customer site but saving a trip. The third option is to deliver to both of them on Tuesday, resulting in increasing the cycle time for the first customer but saving a trip. The example assumes that the combined demand for both customers may be met by one truck.

Inventory consolidation

This approach is related to increasing the order size to benefit from the quantity discounts offered by the vendors. This may also help to take advantage of better transportation rates resulting from higher weight or volume. The downside of the approach is increased inventory carrying costs.

Equipment consolidation

Examples are break-bulk operations of trucking companies and hub-and-spoke operations of airlines. If the demand for origin–destination pairs AD, BD and CD are very small and if A and B are close to C then it may be better to use small trucks to bring in the load from A and B to C and then send them on a big truck to D by combining it with the load from C. Another example is when demands from certain retail stores are small; it may be better to locate one warehouse close to these retail stores, ship in truck loads to the warehouse and ship in small loads to the retail stores.

Let us consider a very simple air transportation network shown in Figure 8.2. The network has three cities on the west coast of the United States (SFO, LAS and LAX) and three cities in the east coast (NYC, ATL and MIA). Considering only one direction, west to east, there are a total of nine origin–destination demand pairs. Assuming that the demands occur over the same period, with no intermediate hubs, nine services (and hence nine separate equipment) are required to fulfill the demand. However, if a hub is located at Dallas (DFW), the number of services can be cut down to three; SFO–DFW–NYC, LAS–DFW–ATL and LAX–DFW–MIA. In addition, it also has the capability now to serve nine

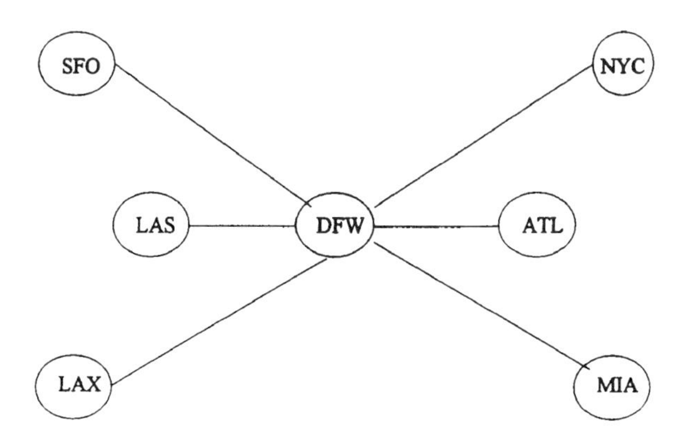

Figure 8.2 Air transportation network: a hub-and-spoke system. Node descriptions refer to US airport codes.

additional origin–destination demand pairs such as SFO–DFW, LAS–DFW and so on. All passengers from different origins but to the same destination will change airplanes in Dallas. For instance, the SFO–DFW–NYC service will carry all passengers from SFO–DFW, SFO–NYC, SFO–ATL and SFO–MIA. Passengers traveling to ATL and MIA will change airplanes at DFW and board the airplane LAS–DFW–ATL and LAX–DFW–MIA, respectively.

In this example the airline could have selected any other city as a hub, or more than one city as hubs. The decision will be based on a trade-off between the costs of having hubs versus not having hubs. The costs of having a hub are the fixed cost of building a hub, the additional handling cost at the hub, the additional variable equipment and crew cost while at the hub, and inventory costs resulting from a longer route for a certain demand. The costs of not having a hub are added equipment and crew costs and equipment underutilization costs. The decision to select the cities to serve as hubs and the assignment of origin–destination demand through one or more hubs can be viewed as a trans-shipment problem. The resulting solution will indicate the cities to serve as hubs and the traffic flow patterns to determine the service.

The modeling and solving of hub location problems involves an understanding of the following issues:

- the number of hubs;
- single or multiple allocation – a demand–supply point may receive–send through one hub or multiple hubs;
- the distance measure may be straight-line, Euclidean or actual distance;
- cost functions may be linear or non-linear;
- the objective may be to minimize the total distance or cost or to minimize the maximum distance;
- constraints such as hub handling capacity and minimum traffic on a spoke or link.

There are a number of models and procedures available in the literature on hub location and planning that address most of the above issues (e.g. O'Kelly, 1986; Campbell, 1994).

Freight consolidation strategies

Several heuristic strategies may be used to accomplish equipment or space consolidation. The three most commonly used strategies are: nearest terminal routing, minimum distance routing and minimum cost routing.

Nearest terminal routing

Shipments are sent to the terminal nearest to the origin or destination or both. All shipments from an origin are sent to the same nearest terminal regardless of destination. This may result in the backtracking of some shipments. No line haul transportation may be involved if the terminal closest to the origin is also the closest terminal to the destination. Nearest terminal routing attempts to minimize local transportation distance and cost; however, it may not minimize the overall distance and cost.

Minimum distance routing

In this case, shipments are sent to the terminals that will minimize the total travel distance. Hence shipments from an origin may be sent to different terminals based on their destination. Since local and line haul rates are different, minimum distance routing may not result in minimum total cost.

Minimum cost routing

Under minimum cost routing, shipments are sent through terminals in order to minimize the total transportation cost. Hence, backtracking may occur if consolidation can lead to lower cost. Again, shipments from the same origin may be sent to different terminals depending upon their destination.

8.4.6 Intermodal transportation

Intermodal transportation is a response to changing marketing and distribution requirements for moving all types of cargo. Widespread recognition of the enormous marketing advantages of very large size carriers, global transport alliances and common marketing and distribution strategies has been influencing the attitude of the freight transportation industry and the industry in general with regard to the concept of intermodalism (Muller, 1989). Some of the advantages of intermodalism include a greater choice of routings, improved services, better pricing and the handling of higher volumes. Use of intermodal containers (containers compatible with two or more modes) has greatly improved the intermodal transfer of general cargo. Reduction of the delay when the cargo reaches the transfer point (a delay which results in an added cost) and minimization of the transfer cost at the modal transfer point are some of the challenges faced by intermodalism. Intermodalism also has to cope with the added burden of minimizing the cost of specialized equipment in the case of trailer on flat car (TOFC) and container on flat car (COFC). Further, as for any other transportation problem, time window constraints and the availability of alternative routings of intermodal transport pose additional challenges.

Intermodal routing issues

There are many factors that must be considered when addressing intermodal transportation routing. These issues have an effect on the transportation carrier as well as the transportation customer. The carrier must consider these factors in order to decide the best price at which to offer his or her services to ensure both a good profit margin and a healthy market share. The customer must choose a carrier based on this cost and his or her desired service level.

Transportation cost

The biggest factor to be considered is the transportation cost. This cost can be based on many factors. It can be a fixed unit cost, a cost per weight unit, a cost per mile, a cost per volume unit or a cost per ton–mile. Transportation cost covers the carriers' fuel costs, equipment cost, crew costs, overhead costs and general and administration costs. When multiple modes are not considered this is the only cost considered while making decisions regarding shipment of goods.

Link-dependent transfer cost

When different modes are considered in transportation a transfer cost is also incurred. A transfer cost is the cost of transferring the cargo from one mode of transportation to another. This may be a relatively simple transfer, as in transferring a trailer from a train's flat bed to the ground so that it can be hooked onto a truck, or it may be a more complex transfer such as the unloading of boxes from a dock and transferring them to an airline terminal to be loaded onto an airplane. This can be thought of as the 'handling' part of 'shipping and handling' charges.

Transfer costs can be either fixed or variable. As in transportation costs these cost can be on a fixed per unit, per weight unit or per volume unit basis. These costs may also be dependent on the city or stop at which they occur as well as on the incoming and outgoing modes. For example, will it cost the same to transfer three crates of fine china in Chicago as it will to transfer them in San Antonio? The difference in cost may be a result of a number of factors such as local labor costs or different equipment needs. Similarly, the cost to transfer from rail to truck may be different from rail to plane at a given city.

Service level and commodity type

Another factor involved in intermodal transportation decisions is that of service level. If service level is of great importance to the customer, transportation by air may be more desirable than by truck. Transfer times must also be considered when evaluating the best method for transportation. Service level is a constraint that must be considered when evaluating the transportation alternatives.

The type of commodity being shipped is another factor that must be considered. The cargo being shipped may have an affect on the best mode or combination of modes of transportation. If an item is perishable the availability of refrigeration and/or the speed of transport might be important considerations. There are also shipping laws and regulations which may need to be considered. There can be commodity–mode compatibility issues which must be investigated when choosing transportation methods.

Modeling complexities

In trying to model intermodal transportation systems there are many components which must be considered. Various factors can affect the representation of reality and the complexity of the model. There is a trade-off: with increased accuracy and reality comes increased model solving time and cost.

Single and multiple routes

One factor to be considered by the modeler is the consideration of single or multiple routes between shipment origin and destination. For example, is there only one way to get from Boston to Seattle or are there many possible paths? In reality there are of course multiple routes, but with each additional path considered there is a large increase in the complexity of the model as there are already a number of modes available at each city to be considered.

Linear and non-linear costs

Linear and non-linear transportation costs add further intermodal transportation modeling complexity. Transportation costs can be modeled in a linear fixed cost per unit (mile, pound weight, etc.) basis. Non-linear transportation costs such as quadratic cost functions or exponential cost functions are more realistic (Jara Diaz, 1982) but are much more difficult to model. Assume that the distance between cities A and B is 100 miles and the distance between cities B and C is 100 miles. In a linear cost model the transportation cost might be $0.20 per mile. A non-linear model might include a cost break at the 100 mile point, where the transportation cost up to and including 100 miles is $0.20 per mile but the per mile cost for distance over 100 miles is $0.15 per mile. So the cost to transport goods from city A to city C directly will be $40 ($0.20 per mile × 200 miles) in the linear cost model. For the same distance, it would cost $30 ($0.15 per mile × 200 miles) in the non-linear cost model. Also, the number of options available to get to the final destinations increases because of non-linear costs.

Shipment splitting

Shipment splitting can also be considered in intermodal transportation modeling. Shipment splitting involves not treating the shipment as one unit. If a shipment is split, on a given segment, a portion of the shipment is shipped via one mode and the remaining portion is shipped via (one or more) other modes. For instance, suppose that 200 lb of cargo are being shipped from Dallas to Chicago; if shipment splitting is implemented 100 lb of the cargo may be shipped via rail, and the remaining 100 lb shipped via truck. The cargo may be separated at the origin or at intermediate points and then reconciled at the destination. This may be done in order to meet weight or volume constraints. Shipment splitting greatly increases the number of possible shipping alternatives. In some cases, shipment splitting may not be allowed, for example when a single piece of a large shipment such as an automobile is being transported.

Multiple objectives

Multiple objectives may also be considered when modeling intermodal transportation decisions. Different objectives may include minimizing total time, minimizing total cost and/or maximizing service level. One possible approach is to weight the different objectives based on their importance. There are several other approaches available to address multiple objectives (Singh *et al.*, 1990).

Fuzzy or imprecise nature of information

The information needed by the carrier in order to determine the various costs or times of transporting goods is often very hard to determine. This imprecise or 'fuzzy' nature of the information can cause the carrier problems in predetermining costs and/or the time of transportation. Transportation time may be affected by weather conditions, traffic conditions or construction of roads or rails. Transfer times and costs may be affected by local labor conditions, equipment availability or weather conditions. Although some shipments may not run into any problems at all, many will at one time or another, making it necessary to plan for uncertainties. Several methods are available to address the fuzzy nature of information (Zimmerman, 1978).

Model for intermodal selection (Reddy and Kasilingam, 1995)

The model uses cost minimization as the objective, assuming that the shipment is not split between modes for a given city pair. The entire quantity is transported with use of only one node between any two cities. Also, transportation costs are assumed to be linear with distance. The notation is as follows:

$c^k_{i,i+1}$ is the transportation cost from city i to city $i+1$ for mode k;
t^{kl}_i is the cost of transferring from mode k to mode l at city i.

The decision variables are as follows:

$$X^k_{i,i+1} = \begin{cases} 1 & \text{if mode } k \text{ is used to transport goods from city } i \\ & \text{to city } i+1, \\ 0 & \text{otherwise;} \end{cases}$$

$$Y^{kl}_i = \begin{cases} 1 & \text{if goods are transferred from mode } k \text{ to mode } l \\ & \text{at city } i, \\ 0 & \text{otherwise.} \end{cases}$$

The model is formulated as follows:

$$Z = \min \sum_i \sum_k X^k_{i,i+1} c^k_{i,i+1} + \sum_i \sum_k \sum_l Y^{kl}_i t^{kl}_i \tag{8.7}$$

subject to the following constraints:

$$\sum_k X^k_{i,i+1} = 1, \forall\, i$$

$$\sum_k \sum_l Y^{kl}_i = 1, \forall\, i$$

$$X^k_{i-1,i} + X^l_{i,i+1} \geq 2Y^{kl}_i, \forall\, i, k, l$$

$$Y^{kl}_i, X^k_{i,i+1} \in \{0, 1\}, \forall\, i, k, l$$

The objective function in this formulation [equation (8.7)] is the sum of two distinct, linear, cost functions. The first cost term on the right-hand side is the sum of total transportation cost for the given quantity to be transported from origin to destination. The second cost term on the right-hand side is the sum of transfer costs, if the shipment is transferred from one mode to another at intermediate cities. The first constraint specifies that only one mode is selected for transporting goods between two cities. This ensures non-splitting of the shipment between modes. The second constraint specifies that only one type of transfer takes place at city i. The third constraint ensures consistency; if at city j there is a modal transfer from k to l then from city i to j the goods are transported by using mode k, and from j to m the goods are transported by using mode l. The final constraint is simply the 0–1 constraint for the decision variables. The optimal solution to this problem would yield a set of zero–one values which would determine the optimal mode for each city pair.

Solution procedure

For m modes and n cities the number of solution alternatives is m^{n-1}. The formulation is very difficult to solve because of the large number of decision variables and constraints and the lack of special structure to it. Hence, a backward dynamic programming approach is proposed to solve the model. The solution procedure is presented below.

Each city is treated as a stage in the dynamic programming procedure. Starting with city $n-1$, the transportation cost plus the transfer costs associated with continuing on the same mode or transferring to a different mode to go to city n is given by:

$$P_{n-1}(k,l) = t^{kl}_{n-1} + qc^{l}_{n-1,n} \quad \forall\, k \tag{8.8}$$

where $P_{n-1}(k,l)$ is the total cost of transportation associated with transferring from mode k to l at city $n-1$, and q is shipment size. Given the incoming mode k, the best choice of outgoing mode at city $n-1$ is m^*, where m^* is determined from the following equation:

$$P_{n-1}(k,m^*) = \min_{l}\{P_{n-1}(k,l)\} \quad \forall\, k \tag{8.9}$$

For cities 2 through $n-2$, (i.e. for $I = 1$ through $n-2$) the corresponding equations to find the best transportation mode are as follows:

$$P_i(k,l) = t^{kl}_{i} + qc^{l}_{i,i+1} + P_{i+1}(l,m^*) \quad \forall\, k \tag{8.10}$$

$$P_i(k,r^*) = \min_{l}\{P_i(k,l)\} \quad \forall\, k \tag{8.11}$$

For city 1, the best mode to go to city 2 is given by the following equation:

$$P_1(s^*) = \min_{k}\{qc^{k}_{1,2} + P_2(k,r^*)\} \tag{8.12}$$

The dynamic programming procedure suggested above will require only $m(n-1)$ calculations, where m is the number of modes between any two city pairs and n is the number of cities on the route. With the above definitions, the step-by-step procedure required to solve an optimal modal combination problem on an intermodal transportation route is as follows:

Step 1: select the best outgoing mode at city $n-1$ by using equations (8.8) and (8.9);
Step 2: select the best outgoing mode for cities 2 through $n-2$ by using equations (8.10) and (8.11);
Step 3: calculate $P_1(s^*)$ by using equation (8.12).

The optimal modal combination is obtained by tracing back through the results of equations (8.12), (8.11) and (8.10). An illustration is provided in Example 8.9.

EXAMPLE 8.9

Consider a transportation route with five cities and three modes of transport (namely, rail, road and air) between any two cities. The transportation costs for the various city pairs are:

Mode	*City pair*			
	1–2	*2–3*	*3–4*	*4–5*
Rail	3	4	3	6
Road	2	4	5	5
Air	4	1	6	4

The intermodal transfer costs are:

Mode	*Transfer cost*
From rail to:	
road	2
air	1
rail	0
From road to:	
road	0
rail	2
air	1
From air to:	
air	0
rail	2
road	1

The shipment size, q, is assumed to be 20 units. Determine the optimal modal combination.

The above example is solved by using the solution methodology presented above. Sample calculations for the four steps are presented below.

Step 1: sample calculations for city 4 with rail as the incoming mode are as follows:

$$\mathrm{P}_4(\text{rail, air}) = t_4^{\text{rail air}} + qc_{4,5}^{\text{air}} = 1 + (20 \times 4) = 81$$
$$\mathrm{P}_4(\text{rail, truck}) = t_4^{\text{rail truck}} + qc_{4,5}^{\text{truck}} = 2 + (20 \times 5) = 102$$
$$\mathrm{P}_4(\text{rail, rail}) = t_4^{\text{rail rail}} + qc_{4,5}^{\text{rail}} = 0 + (20 \times 6) = 120$$

Hence, with rail as the incoming mode at city 4 the best outgoing mode is air. Similar calculations for other incoming modes lead to the best outgoing modes at city 4 and are summarized in the table at the end of this example.

Step 2: sample calculations for city 3 with rail as the incoming mode are as follows:

$$P_3(\text{rail, air}) = t_3^{\text{rail air}} + qc_{3,4}^{\text{air}} + P_4(\text{air, air}) = 1 + (20 \times 6) + 80 = 201$$

$$P_3(\text{rail, truck}) = t_3^{\text{rail truck}} + qc_{3,4}^{\text{truck}} + P_4(\text{truck, air}) = 2 + (20 \times 5) + 81 = 183$$

$$P_3(\text{rail, rail}) = t_3^{\text{rail rail}} + qc_{3,4}^{\text{rail}} + P_4(\text{rail, air}) = 0 + (20 \times 3) + 81 = 141$$

Step 3: from equation (8.12), the calculations for the best mode between cities 1 and 2 are as follows:

$$P_1(\text{rail}) = qc_{1,2}^{\text{truck}} + P_2(\text{rail, air}) = (20 \times 3) + 164 = 224$$

$$P_1(\text{air}) = qc_{1,2}^{\text{air}} + P_2(\text{air, air}) = (20 \times 4) + 163 = 243$$

$$P_1(\text{truck}) = qc_{1,2}^{\text{truck}} + P_2(\text{truck, air}) = (20 \times 2) + 164 = 204$$

The summary of calculations for steps 1–3 are summarized in the following table.

Incoming mode	*Best outgoing mode*	*Total cost*
City 4:		
rail	air	81
air	air	80
truck	air	81
City 3:		
rail	rail	141
air	rail	143
truck	rail	143
City 2:		
rail	air	164
air	air	163
truck	air	164

The optimal solution to the problem is obtained as follows. The best mode from city 1 to city 2 is truck, as seen from step 3. Given that the incoming mode at city 2 is truck, the best mode from city 2 to city 3 is air. Given that the incoming mode at city 3 is air, the best mode from city 3 to city 4 is rail. Given that the incoming mode at city 4 is rail, the best mode from city 4 to city 5 is air. The optimal modal combination is therefore:

City pair	1–2	2–3	3–4	4–5
Mode	truck	air	rail	air

8.5 ADVANCED TRANSPORTATION MODELS

In this section four transportation problems that are specific to certain types of industry are presented. All of these models are very large and complex in terms of size and the type of constraints incorporated. The first addresses air cargo revenue management. It describes the nature of the problem and the models and procedures used to solve the problem. The second is in the area of fleet sizing for a retail company involved in managing its own transportation fleet and operations. The last two problems are from the rail road industry. The first of these addresses the management and planning of a very critical resource – locomotives. The second addresses one of the dimensions of the rail road operating plan – the blocking plan.

8.5.1 Cargo revenue management (Kasilingam and Hendricks, 1993)

Yield management, alternatively known as revenue management, can be defined as the integrated management of price and inventory to maximize the profitability of a company. In the case of combination air carriers, or airlines operating airplanes that carry passengers as well as cargo, it is the management of passenger fares and seats together with cargo rates and belly space. In practice, the concepts of airline yield management are applied in a hierarchical manner. For instance, an airplane can take off with a certain weight (payload) which is available for carrying passengers, passenger bags and cargo. The main cabin has a fixed number of seats available for carrying passengers, and some overhead cabin space for carry-on bags. The belly has space available for accommodating passenger bags and cargo. The number of seats available for sale is managed by a passenger yield management (PYM) system.

The weight available for cargo sales is equal to payload less the weight of the expected number of passengers on board, based on PYM controls and passenger bookings and their bags. The belly volume available for cargo sales is equal to the total belly space less the space of the passenger bags. In the case of a wide body airplane, the number of positions available for cargo sales is equal to the total number of positions (containers) in the belly less the number of containers required for passenger bags. The weight and volume and positions available for cargo is managed by a cargo revenue management (CRM) system. If mail is not treated as cargo and has a higher boarding priority than has other cargo then the weight and volume or positions occupied by mail should be subtracted when computing the space available for cargo.

With substantially low passenger yields and relatively saturated passenger traffic the ability of cargo to provide additional revenue is becoming more and more important these days. Industry forecasts predict a tremendous growth in cargo demand; worldwide cargo volume is expected to double by the year 2005. Hence it is becoming more and more important to manage the available cargo space effectively (weight and volume and positions). The objective here is to highlight the major differences between a PYM system and a CRM system and to discuss some of the complexities involved in developing and implementing a CRM.

Passenger yield management versus cargo revenue management

Cargo yield management differs from passenger yield management in many respects. Four important differences are discussed here.

Uncertain capacity

PYM controls a fixed and known number of seats. In CRM the weight and volume and positions available for sale are not fixed; they depend on the payload, belly space and the expected number of passengers on board and their bags. In addition to the variability of the expected number of passengers on board, payload is also a variable. It depends upon several factors such as runways, weather, fuel weight, ramp weight etc. This introduces the necessity to develop models to forecast the capacity available for cargo sale. This in turn makes one of the key inputs of the overbooking model – capacity – stochastic in nature. In PYM overbooking models, capacity is assumed to be known and deterministic.

Three-dimensional capacity

Cargo capacity is three-dimensional (weight, volume and number of container positions). For instance, when booking a low-density shipment, capacity may be available in terms of weight but not in terms of volume. Sometimes, weight and volume may be available to accommodate the shipments, but they may not fit inside a container because of their different shapes. This results in what is known as stacking loss. The three-dimensional nature of capacity necessitates the need to work with weight as well as volume or position capacity forecasts. This can sometimes be addressed by using standard weight–volume relationships or density values established from historical data.

Itinerary control

Passengers prefer to follow their planned itinerary without being bumped or re-routed. On the other hand, cargo may be shipped along any route as long as it is available at the destination within the specified or agreed upon delivery date and time. Hence, multiple routings may be available to ship cargo from origin to destination. This adds one more dimension to the traditional PYM capacity or bucket allocation models used to allocate space to different fare classes or services.

Allotments

For most carriers 'allotments' take up a major chunk of the cargo space. Allotments are space reserved for big customers (major shippers and forwarders) on certain flights for certain days of the week over a period of time. This space is not available for general sale. This practice requires decision support to identify the flights for which allotments need to be set up and the amount of allotment space based on stochastic cargo capacity as well as the displacement and demand effects of general cargo.

The above key differences warrant a special type of yield management system with models more complex than a traditional PYM system. For instance, mathematically speaking, the overbooking model should be able to handle the stochastic nature of cargo capacity. The allocation model needs to address multiple routings between an origin and a destination. An additional model is required for setting up allotments. Also, the relationship between weight and volume is another important issue to be addressed in all the yield management models.

Cargo revenue management process

The cargo revenue management process consists of four important steps. The first step is to forecast the cargo capacity available for sale, in terms of weight, volume and positions. This step is not required in a PYM system since capacity is known and deterministic. The second step is to allocate space for 'allotments' based on the demand and profitability from allotments and from anticipated general sale. The third step is to overbook the remaining capacity, the forecasted capacity less allotments, to compensate for cargo booking behavior in terms of cancellations, no-shows and variable tendering. The final step is to allocate the overbooked capacity to different markets and products in a way that maximizes the total cargo revenue or profitability. This is known as bucket allocation or capacity allocation. The last two steps are the same as in PYM; however, the models and procedures used in CRM are different.

Several other supporting models similar to that in PYM are required besides capacity forecasting, allotment determination, overbooking and bucket allocation, for instance forecasting of cargo demand by market and product class, forecasting of no-shows, cancellations and variable tender behavior, computing oversale and spoilage costs and determining nesting mechanisms (note: variable tendering refers to the act of showing up with less or more cargo than that was booked). Some of the above models are related to the planning aspects of the revenue management system, others are related to the reservation control and booking aspects of the revenue management system.

The CRM process is performed on a nightly basis, very similar to the PYM process. The first control decisions for a flight are typically made 8–10 days before departure. Thereafter, the decisions are updated almost daily until the flight departs. Days at which the decisions are updated are known as 'reading days'. For instance, reading day 2 is two days before flight departure. Additional information available (cancellations and new bookings) for the flight between reading days and post-departure information from similar flights (that departed to the same destination around the same time on the same day of the week) such as cargo capacity and cargo on board are used in updating the decisions.

Pre-departure flight management, generation of periodic and *ad hoc* management reports and performance monitoring of the forecasting and optimization models are an intrinsic part of a CRM as well. These issues are not addressed in here since they are pretty much similar to their counterparts in PYM. The two reports that are specific to cargo are the allotment usage report and the load factor report. The allotment usage report provides information on the utilization of the allotted space by flight and by customer over a specified period of time. The load factor report in cargo differs from the passenger side because of the three-dimensional nature of the cargo capacity. Load factors in terms of weight, volume and positions are reported for wide body flights. For narrow body flights, volume and weight load factors are reported. The scope of this section is to address the complexities involved in the forecasting and optimization decision support models.

Cargo revenue management models

The relationship among the different models in CRM is shown in Figure 8.3. The important models are discussed in detail in the following subsections.

Capacity forecasting

Forecasting capacity is the first step in a CRM system. Capacity should be forecasted in terms of weight, volume and position. The following equations define cargo capacity in all three dimensions. These equations assume that mail has a higher boarding priority than has general cargo. For narrow body flights, the weight and volume capacities are defined as follows:

$$\text{weight capacity} = (\text{payload}) - (\text{passenger weight}) - (\text{bag weight}) \\ - (\text{mail weight}) - (\text{extra fuel weight})$$

where extra fuel weight is the weight of the additional fuel required to carry cargo;

$$\text{volume capacity} = (\text{belly volume}) \times (1 - \text{stacking loss}) - (\text{bag volume}) - (\text{mail volume})$$

For wide body flights, we have:

$$\text{weight capacity} = (\text{payload}) - (\text{passenger weight}) - (\text{bag weight}) - (\text{mail weight}) - (\text{extra fuel weight})$$

$$\text{position capacity} = (\text{total positions}) - (\text{bag positions}) - (\text{mail positions})$$

$$\text{volume capacity} = (\text{position capacity}) \times (\text{volume per position}) \times (1 - \text{stacking loss})$$

Operationally, the capacity available depends upon the density of the booked shipments and their shapes. Expected passengers on board for a flight is obtained from the PYM system. Standards for passenger weight, bag weight, bag volume, freight density and stacking loss are usually established from industrial engineering studies. Historical data are used in a few cases to supplement these studies. Payload and mail weight are typically forecasted from historical data. This may be done by using traditional exponential smoothing models with appropriate trend factors and seasonality indices. Most airlines have historical payload data. By analysing the historical payload patterns appropriate

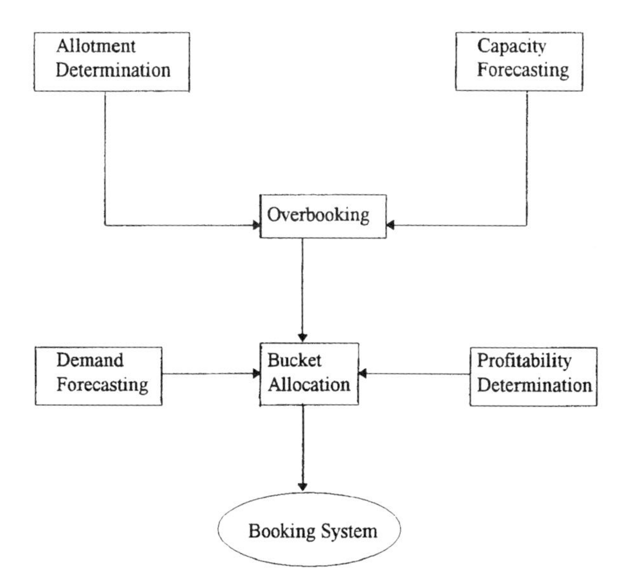

Figure 8.3 Cargo revenue management models.

exponential smoothing models and model parameters may be selected. Payload can also be forecasted by means of causal models. Since payload depends on temperature, wind, ramp weight, fuel weight and a few other factors regression-based causal modeling is another possibility. The selection of a suitable approach, or method, will be based on the performance of the forecasts in terms of bias and accuracy over a given period of time.

Allotment determination

Determination of allotments for different customers requires careful consideration of a lot of important factors. It must be made sure that space is available on the flight to accommodate the allotment request. Of course, the type of container the customer is planning to bring in and the flight equipment must be compatible. After confirming space availability and container–equipment compatibility an evaluation must be made about the profitability of the allotment request. This evaluation will be based on the expected profitability of this request versus the expected profitability from other requests and 'general sale' demand.

Modeling overbooking

Overbooking is the practice of intentionally selling more cargo space to compensate for no-shows, cancellations and variable tendering. Overbooking determines the capacity available for sale. Variability may be in terms of weight and/or volume. The overbooking model considers the costs of oversale and spoilage along with the mean and variance of show-up rate. Show-up rate is a composite index computed based on no-shows, cancellations and variable tendering. In addition to these factors, cargo overbooking must address the stochastic nature of capacity. This adds to the complexity of the cargo overbooking model in terms of its formulation and solution procedure. Uncertainty in capacity forecasts is modeled by considering the mean and variance of capacity in the form of a distribution.

The objective of overbooking may be to minimize the sum of oversale and spoilage costs or to maximize the expected net revenue, which is equal to the total revenue minus the expected oversale cost. Sometimes it may be desirable to add a constraint on the allowable oversales to limit the 'economic' overbooking level considering the practical aspects of oversales. The mathematical expressions for total revenue, expected oversales and expected spoilage are derived by means of probability theory. These expressions and the difficulty of deriving them depend to a large extent on the underlying assumptions of the distributions for capacity and show-up rate. An overbooking model based on minimizing the sum of oversale and spoilage costs is given below (Oakley *et. al.*, 1992). The notation used is as follows:

B is the show-up rate, with a probability density function (PDF) $f(B)$;
C is the capacity with a discrete or continuous probability distribution, with PDF $h(C)$;
V is the oversale cost per unit of capacity;
R is the spoilage cost per unit of capacity;
L is the overbooking level.

In its simplest form, assuming all values are deterministic, the overbooking level may be computed as given below. Assuming reservations book to the overbooking level, L is equal to C/B. When B is less than C/L, spoilage occurs. The amount of spoilage is equal to $C - BL$ and the corresponding spoilage cost is equal to $R(C - BL)$. When B is greater than C/L, oversale occurs. The amount of oversale is equal to $LB - C$ and the corresponding oversale cost is equal to $V(LB - C)$. Suppose we relax the assumption that the show-up rate is deterministic. Then the expected spoilage costs, E^{SC}, and oversale cost, E^{OC}, are as follows:

$$E^{\mathrm{SC}} = R \int_0^{C/L} (C - LB)\, f(B)\, \mathrm{d}B \tag{8.13}$$

$$E^{\mathrm{OC}} = V \int_{C/L}^{1} (LB - C)\, f(B)\, \mathrm{d}B \tag{8.14}$$

Further, if we relax the assumption that capacity is deterministic and assume that C follows a discrete probability distribution, taking on values $C_1, C_2, \ldots, C_n$ with probabilities $p_1, p_2, \ldots, p_n$, and if $\sum p_i = 1$, then the expected total cost, E^{TC}, is given by:

$$E^{\mathrm{TC}} = E^{\mathrm{SC}} + E^{\mathrm{OC}} \tag{8.15}$$

$$E^{\mathrm{TC}} = \sum p_i \left[R \int_0^{C_i/L} (C_i - LB)\, f(B)\, \mathrm{d}B + V \int_{C_i/L}^{1} (LB - C_i)\, f(B)\, \mathrm{d}B \right] \tag{8.16}$$

Differentiating with respect to L and setting it equal to zero, we obtain

$$\sum p_i \left[V \int_{C_i/L}^{1} B\, f(B)\, \mathrm{d}B - R \int_0^{C_i/L} B\, f(B)\, \mathrm{d}B \right] = 0 \tag{8.17}$$

Let $\mu_B = \int_0^1 B\, f(B)\, \mathrm{d}B$; then equation (8.17) becomes:

$$\sum p_i \left[V\mu_B - (V + R) \int_0^{C_i/L} B\, f(B)\, \mathrm{d}B \right] = 0 \tag{8.18}$$

$$V\mu_B \sum p_i - (V + R) \sum p_i \int_0^{C_i/L} B\, f(B)\, \mathrm{d}B = 0 \tag{8.19}$$

Or, since $\sum_i p_i = 1$,

EXAMPLE 8.10

Assume $V = R = \$10.00$ per unit of inventory. Suppose the show-up rate is distributed uniformly over $[0.8, 1.0]$ and that the PDF for capacity, $P(C = c)$, is as given below:

Value	100	110	120
$P(C = c)$	$\frac{2}{5}$	$\frac{2}{5}$	$\frac{1}{5}$

$f(B) = 5$, $0.8 < B < 1.0$, and $\mu_B = 0.9$

From equation (8.20), we obtain:

$$\frac{2}{5}\int_{0.8}^{100/L} 5B\,\mathrm{d}B + \frac{2}{5}\int_{0.8}^{110/L} 5B\,\mathrm{d}B + \frac{1}{5}\int_{0.8}^{120/L} 5B\,\mathrm{d}B = \frac{10 \times 0.9}{10 + 10}$$

This leads to $29\,300/L^2 = 2.05$, and thus $L = 119.55$. Evaluation of E^{TC} at $L = 119$ and at $L = 120$ with use of equation (8.16) gives E^{TC} values of 83.29 and 83.33, respectively, resulting in an optimal value of L of 119.

$$\sum_i p_i \int_0^{C_i/L} B\,f(B)\,\mathrm{d}B = \frac{V\mu_B}{V + R} \tag{8.20}$$

The optimal overbooking level can be obtained by solving equation (8.20). Depending upon the type of discrete distribution, it may be solved analytically or numerically. An illustration is provided in Example 8.10.

Modeling bucket allocation

Bucket allocation is the process of determining the space to be allocated to different markets and products on a given flight based on their demand and profitability. Cargo products are usually defined in terms of service patterns such as priority (same-day, next-day and general). The objective of bucket allocation may be to maximize system profitability subject to certain constraints. These constraints are related to the overbooked capacity available for sale and the demand for various markets and fare classes. Bucket allocation provides the necessary mechanism to manage cargo bookings by origin–destination pairs and products.

In PYM, passengers usually decide on their itinerary while booking and fly according to their booked itinerary. In the case of cargo, demand is at the market level, and routing is flexible as long as the service time commitment is met. Routing flexibility may also include the options of using other carriers as well as using other transportation modes on certain legs. This provides flexibility operationally but adds one more dimension to the complexity of the allocation model in terms of multiple routings between an origin and destination. Also, the model should be formulated to address the network effects of long-haul markets displacing local markets and of redirecting demand from low-load factor flights to high-load factor flights. A formulation of the bucket allocation problem to maximize the total system contribution, assuming normally distributed demand, is presented below. The notation is as follows:

i is the index of market product, $i = 1, \ldots, N$;
j is the index of routing, $j = 1, \ldots, J_i$;
k is the index of flight k, $k = 1, \ldots, K$.

The decision variable is as follows:

X_{ij} is the allocation for market or product i using routing j.

The parameters are as follows:

$$A_{jk} = \begin{cases} 1 & \text{if flight } k \text{ is part of routing } j, \\ 0 & \text{otherwise;} \end{cases}$$

C_{ij} is the unit variable route cost for shipping market or product i along route j;
D_i is the demand for product i and demand is assumed to follow a normal distribution with parameters (μ_i, σ_i);
R_i is the unit revenue from selling market or product i;
T_k is the authorized cargo capacity of flight k;
W_{ij} is the unit contribution from selling market or product i and using routing j (equal to $R_i - C_{ij}$).

Note: allocation for market or product i on flight k is obtained from $\sum_j X_{ij} A_{jk}$.

The objective function of the model is to maximize the total contribution (revenue minus variable cost) from all the markets or products over all flights in the system:

$$Z = \text{maximize} \sum_i \sum_j W_{ij} X_{ij} \tag{8.21}$$

The flight leg capacity constraints are modeled as follows:

$$\sum_i \sum_j X_{ij} A_{jk} \leq T_k \quad \forall\, k$$

The uncertainty in demand is modeled as a chance constraint. The assumption here is that the management requires the probability of allocation for a market or product from all routings exceeding the demand to be less than or equal to α_i. Hence, we have

$$P\left[D_i \leq \sum_j X_{ij}\right] \leq \alpha_i \quad \forall\, i$$

For normally distributed demand it can be shown that the above is equivalent to the following expression, where Z_{α_i} is the standard normal variate such that:

$$\sum_j X_{ij} \leq \mu_i + Z_{\alpha_i}\sigma_i, \quad \forall\, i$$

The final constraint concerns integrality restrictions on the decision variables and is represented as follows:

X_{ij} is an integer for all i and j

Modeling of the bucket allocation problem typically results in non-linear, stochastic programming formulations. Their complexity depends on the underlying assumptions of the probability distribution for demand. It is very hard to develop procedures to solve these formulations exactly. A simple, alternative, approach is to model the bucket allocation problem as a deterministic formulation and solve it using Lagrangian relaxation-based techniques (Fisher, 1981). Relaxing the demand constraints decomposes the problem into N subproblems, one for each market or product.

Demand forecasting

Demand forecast for different cargo products at the market level is one of the key inputs for bucket allocation. Forecasting demand for various cargo products for all markets is done by combining long-term and short-term forecasts by means of appropriate techniques. Long-term forecasts are obtained by using exponential smoothing models based on historical data. Short-term forecasts are obtained from pre-departure booking information and historical booking profiles. Unlike passenger demand, most cargo demand occurs in a very short time window close to departure. Also, the number of bookings of cargo is usually very small. Added to this, the sizes of bookings vary significantly. This necessitates different approaches to forecasting such as using the demand forecast for a few major customers as a basis for market forecast and incorporating causal factors for certain markets during certain periods. Forecasting of no-shows, variable tendering and cancellations is done very similarly to demand forecasting using appropriate data.

Cargo profitability

In the case of passengers it is sufficient to define profitability simply based on revenue and costs. In the cargo business profitability should be defined at a global level. For instance, a customer may not be profitable in one segment; however, he or she may be providing a lot of business to the airline and the overall system profitability from the customer may be very high. It may not be wise to displace a regular, profitable shipper in favor of a one-time shipper with a high profitability. Also, the show-up behavior of shippers is another important criterion. Some shippers may promise a lot of business but at departure may bring only very little to ship. Hence, computation of cargo profitability must include shipper behavior, business volume and profitability both at the segment level as well as at the system level.

8.5.2 Trailer management system (Kasilingam and Ong, 1995)

Management of the inventory and movement of trailers is a very challenging operational problem in the trucking and retail industry. Non-availability of empty trailers at the right location at the right quantity will delay or cancel the shipment of merchandise to the customers or stores. It is also not wise to maintain a large inventory of trailers because of their high cost. The issue of trailer/tractor ratio has a significant relationship with a host of other factors such as customer service levels, warehouse capacity, vehicle dispatching rule and handling and storage methods at distribution centers, vendor sites and stores.

Functional requirements of a trailer management system

A trailer management system (TMS) should:

- operate as a planning tool to estimate the long-term trailer requirements on a monthly basis; this estimation should account for the seasonal aspects of business and renting options while generating the monthly trailer requirements;
- have the capability to be used as an operational tool to determine the trailer requirements at a warehouse or store based on the current availability of trailers at different locations and the demand for merchandise movement;
- have the capability to perform what-if analysis to test different scenarios; the scenarios may include relocation of warehouses or stores, growth aspects such as addition of stores or warehouses, elimination of store drop programs as well as change in dispatching methods.

A good TMS should consist of decision models, databases and a graphical user interface. It should provide the dispatch analysts with models for long-term and operational planning of trailers as well as tools for interactive query capabilities.

Features of a trailer management system

A good decision support system for trailer management will have the following features:

- long-term trailer planning capabilities considering seasonal variations and renting;
- day-to-day operational planning of trailer movement to meet the demand for merchandise movement;
- ability interactively to run trailer management models, change model inputs and review and modify model outputs; this will include running *ad hoc* queries and performing what-if and sensitivity analysis;
- creation of management reports on trailer utilization, movement and requirements;
- ability to manage trailers proactively; the system will identify warehouses or stores that require attention based on user-defined critical parameters and will alert the dispatch analysts.

Functional overview of trailer management systems

An ideal decision support system for trailer management will have four major components: a demand forecasting model to determine the number of trailer loads by origin and destination; a heuristic model to plan long-term trailer requirements and to determine the number of trailers to be rented to meet peak requirements; a simulation model to perform what-if analysis and operational planning; and, finally, a database to support all the models. Figure 8.4 shows the relationships among the components of the TMS.

Demand forecasting model
Demand for trailer loads at the origin–destination level are required to determine the number of trailers required. The demand for trailer loads is not always known *a priori* with certainty. Hence, a forecasting model is required to generate origin–destination forecasts of trailer loads. The forecasting model will use historical data to produce forecasts. An appropriate forecasting model and model parameters will be selected based on historical demand patterns to result in the best forecasting accuracy. Selection of the model will consider seasonal and trend aspects of trailer loads. The model will also employ suitable methods of pre-processing data, which will include the screening of outliers and the transformation of demand data. The output from the forecasting model will be used by the heuristic and simulation models.

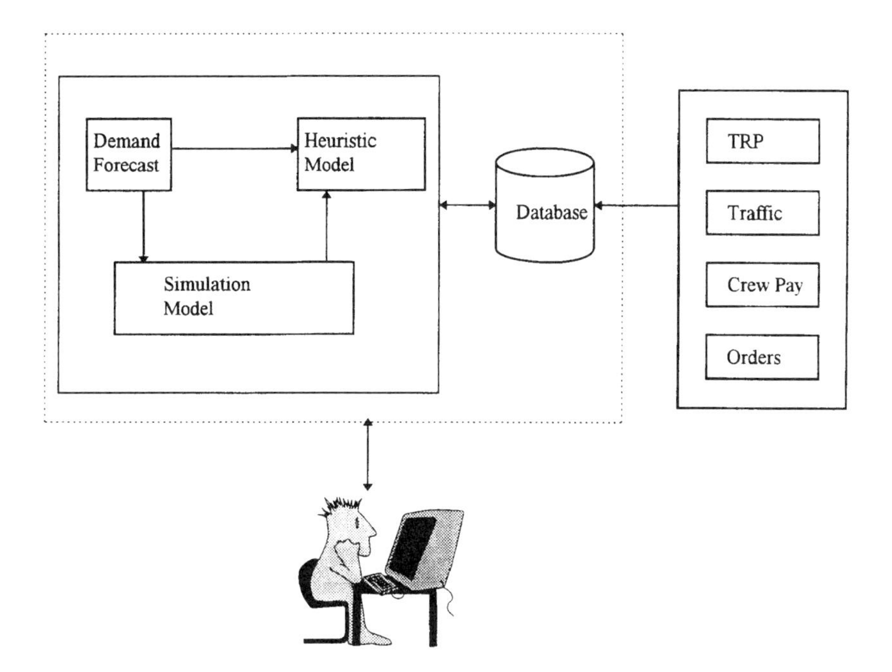

Figure 8.4 Components of a trailer management system. TRP = transportation resource plan data.

Heuristic model

This is one of the core components of the TMS and is essentially a decision model to estimate long-term trailer requirements. The model will determine the number of trailers to be owned over a period of time and the additional trailers to be rented to meet seasonal demand. Sensitivity analysis can also be performed with respect to some of the model inputs to see the effect of change in inputs on the trailer requirements. For instance, an increase in trailer rental cost may reduce the number of trailers to be rented. The heuristic model can also be used to generate good alternative solutions for subsequent analysis by the simulation model. The model will reduce the number of trailers through optimal combination of ownership and rental to meet the current service level. The objective of the heuristic model will be to minimize the cost of trailer ownership (purchasing and maintenance costs), renting cost and line haul mileage cost for rented and owned trailers. The model will have two major constraints. The first constraint will ensure that the trailer loads of demand between various locations are met. The second constraint will ensure that a trailer is not used for more than the available time in a time period. The model is expected to be very large in terms of the number of variables and constraints. Specialized procedures should be developed to solve the model.

Simulation model

The simulation model can be used both as an evaluation tool and as a prescriptive tool. As a prescriptive tool it can be used to determine long-term trailer requirements and to determine trailer requirements at the operational level, considering real-time constraints. As a long-term planning tool the simulation model is expected to provide more accurate trailer requirements than the heuristic model since it considers operational issues such as dispatching rules and arrival and departure patterns. However, it is more expensive in terms of computer time than is the heuristic model. Also, whereas a heuristic model determines the trailer requirements with use of costs and other constraints, a simulation model requires a set of alternative trailer counts to be input for evaluation. A simple empirical procedure can be designed to generate 'good' alternative trailer counts or the solution from the heuristic model can be used as a starting point.

As an operational planning tool, it can be used to identify warehouses or stores that require attention based on some user-defined critical parameters. For instance, the simulation model will use the current situation as an input to analyze trailer flow through the system and will determine the trailer requirements at the various locations for the next 24 hours. The output will identify trailer shortages and surpluses at the various locations. If the shortage or surplus exceeds a critical value dispatch will be alerted. This will enable the dispatch analysts to manage trailer requirements proactively. As an evaluation tool the simulation model can be used to analyse certain scenarios, for instance the impact of eliminating the 'store drop program' on costs, trailer and driver utilization and service levels. Evaluation of a scenario or long-term trailer planning involves several simulation runs based on certain arrival and departure distributions, loading and unloading times, dispatching rules and costs to evaluate the scenarios based on certain user-specified criteria. The use of the simulation model as an evaluation tool requires careful design of the simulation experiment, giving due consideration to experimental design, input data stream preparation and simulation parameters such as run length and number of runs.

Database

The database will contain all the input data requirements for the three models – the heuristic model, the simulation model and the demand forecasting model. It will also contain the outputs generated by the models. The outputs from some of the models are required as inputs to other models and are also required for generating periodic management reports. The data requirements for the demand forecasting model include historical trailer movement between warehouses, stores, vendors and other locations. This may be obtained from driver log sheets and/or from dispatch log sheets. The data requirements for the heuristic model include demand forecasts from the forecasting model, the distance between locations, the maximum distance that can be traveled in a day and cost data such as maintenance and purchase costs of trailers, trailer rental costs and line haul mileage costs for owned and rented trailers. The additional data requirements for the simulation model will include arrival and departure patterns of trailers at various locations, loading and unloading times, crew cost components, service level considerations and data from transportation resource plan (TRP) on dispatches.

8.5.3 Locomotive distribution planning

Optimal assignment of locomotives to trains is very important to railroads because of the high cost of operating locomotives and the impact of locomotives on departure and arrival performance of trains, which in turn affects customer satisfaction. The problem is to determine the optimal number of locomotives to pull trains using various types of available locomotives taking into account the fleet composition, the maintenance constraints and operating costs. The objective is to minimize the operating costs (including total locomotive fleet requirements) and to minimize the train setback (delays). The motive power required to pull a train is determined based on train tonnage and train route. The force required to pull a train is often expressed in terms of horsepower rather than the number of locomotives. The motive power demand for a particular train may be met by using a set of locomotives. Each such set is called a 'consist'. There may be certain restrictions or rules for building a consist: minimum and maximum number of units on the consist, respecting the weight to power ratio, maximum number of deadheading locomotives, incompatibility of locomotives etc. Since the traffic flow is not symmetric the system must handle the repositioning of the units using deadheading locomotives on trains and by making light engine moves where a locomotive is moved between two terminals independently of the trains using a crew.

Locomotive distribution model (Zirati et al., *1997)*

In this section a simplified formulation of the locomotive distribution problem is presented. In this formulation, information on the maintenance constraints and on the outpost work are not presented. Also, only utilization costs are used; no cost on service reliability and no bonus and/or penalty are used to represent more realistic operating conditions. The problem is a multicommodity flow problem with resource constraints. The only resource variable in the problem is the time between the units that need to be routed to the shop. Each commodity represents a locomotive class (each locomotive that must be routed to a shop is considered a locomotive class by itself) and the flow on each arc represents the number of locomotives. A time–space network is constructed for each commodity. The notation is as follows:

K is the set of commodities;
K^c is the number of locomotives that have to be routed to a shop;
$G^k = (V^k, A^k)$ is the network associated with $k \in K$, where $V^k = N^k$ $\{o(k), d(k)\}$ is the node set and A^k is the arc set;
N^k is the number of sites that may be visited by locomotives of class k leaving from the source node $o(k)$ and arriving at the sink node $d(k)$; the arc $(o(k), d(k))$ always exists, to permit conservation of flow when not all locomotives of class k are used;
w is a train (in a set of trains W) that must be covered by the locomotives; the set W is divided into three sets W^n, W^p and W^q containing, respectively, the trains for which motive power requirements are expressed using the minimum number of locomotives, the weight-to-power ratio and tonnage;
n_w is the minimum number of locomotives necessary to cover train task $w \in W^n$;
p_w is the required horsepower for train $w \in W^p$;
c_{ij}^k is the cost of a locomotive of class k pulling tonnage on arc (i, j);
d_{ij}^k is the deadhead cost of a locomotive of class k on arc (i, j);
D_k is the available number of locomotives of class k at the source node at the beginning of the week;
p^k is the operating horsepower of locomotive class k;
q_{ij}^k is the total weight that may be pulled by a locomotive of class k on arc (i, j);
q_w is the tonnage of train $w \in W^q$ to be pulled by locomotives;
T_i^k is the accumulated operating time of the unit $k \in K^c$ at node i, starting from its initial departure at the beginning of the week;
t_{ij}^k is the travel time from node i to node j, $(i, j) \in A^k$ for locomotive of class $k \in K$;

$$a_{w,ij}^k = \begin{cases} 1 & \text{if arc } (i,j) \in A^k \text{ covers task } w, \\ 0 & \text{otherwise.} \end{cases}$$

The decision variables are as follows:

X_{ij}^k is the number of pulling locomotives of class k covering the train task arc (i, j);
F_{ij}^k is the number of deadheads in the consist that covers the train task arc (i, j).

The model that covers locomotive distribution planning for a one-week period is expressed as follows:

$$\min \sum_{k \in K} \sum_{(i,j) \in A^k} (c_{ij}^k X_{ij}^k + d_{ij}^k F_{ij}^k) \tag{8.22}$$

subject to

$$\sum_{k\in K}\sum_{(i,j)\in A^k} a^k_{w,ij}X^k_{ij} \geq n_w, \forall\, w \in W^n$$

$$\sum_{k\in K}\sum_{(i,j)\in A^k} a^k_{w,ij}p^kX^k_{ij} \geq p_w, \forall\, w \in W^p$$

$$\sum_{k\in K}\sum_{(i,j)\in A^k} a^k_{w,ij}q^k_{ij}X^k_{ij} \geq q_w, \forall\, w \in W^p$$

$$\sum_{j:(o(k),j)\in A^k} X^k_{o(k),i} = D_k, \forall\, k \in K$$

$$\sum_{j:(i,j)\in A^k} (X^k_{i,j} + F^k_{ij}) - \sum_{j:(j,i)\in A^k} (X^k_{ji} + F^k_{ji}) = 0, \forall\, k \in K, \forall\, i \in N^k$$

$$\sum_{j:(j,d(k))\in A^k} X^k_{j,o(k)} = D_k, \forall\, k \in K$$

$$X^k_{ij}(T^k_i + t^k_{ij} - T^k_j) \leq 0, \forall k \in K^c, \forall\, (i,j) \in A^k$$

$$a^k_i \leq T^k_i \leq b^k_i, \forall\, k \in K^c, \forall\, i \in V^k$$

$$X^k_{i,j} \geq 0, F^k_{i,j} \geq 0 \text{ and are integers}, \forall\, k \in K, \forall\, (i,j) \in A^k$$

The objective function [equation (8.22)] minimizes the total cost of pulling and deadheading. The train coverage requirements expressed in terms of minimum number of locomotives, horsepower and tonnage are represented by the first three constraints. The fourth and fifth constraints require the transfer of D_k locomotives from origin $o(k)$ to destination $d(k)$ for each $k \in K$. If a locomotive of class k is not required for a task it takes arc $(o(k), d(k))$ to the destination node. Flow conservation at each node is represented by fifth constraint. Note that a locomotive could change status at each node, for example a deadheading locomotive could become active (pulling tonnage). The seventh and eighth constraints are specific constraints for locomotives that have to be routed to a shop. The former is the compatibility constraint between the flow and time variables, the latter is the time windows constraint for units to be routed to the shop. The last constraint guarantees that the flow variables take integral values.

Solution procedure

In the discussion below, a column represents a route of an unspecified locomotive during the weekly period under study. The rows are the constraints that the routes have to satisfy to get a valid solution [e.g. the train coverage constraints (rules to validate if the train demand is satisfied)]. The formulation presented above will result in a very large number of columns and rows for any of the major railroads. It is to be noted that the objective function [equation (8.22)] and constraints are separable by commodity. Since the formulation has a block structure with linking constraints, a natural solution approach is the Dantzig–Wolfe decomposition. The formulation can be decomposed into a master problem and subproblems. The master problem includes all constraints involving more than one commodity (the first three constraints); there is one subproblem (containing the final six constraints) per commodity. A column-generation methodology based on the Dantzig–Wolfe decomposition can be used to solve the master problem. The subproblems may be solved by using a shortest path algorithm. To reduce the problem further it is also possible to use the Dantzig–Wolfe decomposition with other forms of decomposition such as time window. However, it does not guarantee optimality, but experience shows if reasonably large time windows are used the solution is very near to optimal.

8.5.4 Railroad blocking problem

In rail freight transportation general merchandise freight may pass through many classification yards on its route from origin to destination. To realize the economies of scale railroads reclassify inbound traffic from various origins in the yards and put them on outbound trains with the same or close destinations. Reclassification requires labor and yard resources, and it often causes significant delay to the shipment. In order to minimize the number of reclassifications, railroads group shipments going to different destinations into a block, which will not be reclassified until arriving at the destination yard of the block. Several cars with different destinations are grouped together in a block, and cars within a block are not reclassified until the block destination is reached. The objective of the blocking problem is to design a blocking plan that minimizes total reclassification costs. A blocking plan specifies the blocks to be built at each yard and the traffic assigned to the blocks.

Blocking plan example

Example 8.11 provides an understanding of a railroad blocking plan.

Blocking model (Barnhart and Jin, 1997)

A majority of the railroads develop blocking plans by refining existing plans. These refinements are inherently local in nature, which may forego potential improvement that requires tremendous changes to existing plans. Research shows that there is a significant potential to

EXAMPLE 8.11 (NEWTON AND VANCE, 1996)

Consider a railroad with four terminals, A, B, C and D, and three origin–destination (OD) demand pairs. The demands for AB, AC and AD are 100, 80 and 90 cars, respectively. The reclassification capacities at terminals A, B and C are 270, 90 and 90 cars, respectively. Consider the four blocking plans given in the figure. The first plan sends all cars to terminal B; the string of cars may have the OD pairs mixed in some order. At terminal B, OD pair AB has reached its destination. Now OD pairs AC and AD are blocked to move to terminal C. At C, AC has reached the terminating point and OD pair AD is blocked to terminal D which is its destination. Under this blocking plan, 100 cars of OD pair AB use one block, 80 cars of OD pair AC use two blocks with an intermediate reclassification at B and 90 cars of OD pair AD use three blocks with intermediate reclassifications at B and C. Hence, blocking plan 1 results in $260 = [80 + (2 \times 90)]$ intermediate reclassifications. However, blocking plan 1 is not feasible since it requires the reclassification of 170 cars (80 cars for OD pair AC and 90 cars for OD pair AD) at terminal B which has a capacity of only 90 cars. Similar calculations can be performed for other blocking plans. Blocking plan 1 has all short blocks since block destination is always the next terminal. Blocking plan 4 has all direct blocks; in other words, the block destination is the same as the OD pair destination for all OD pairs. In this example, blocking plan 2 is optimal, with 80 intermediate reclassifications.

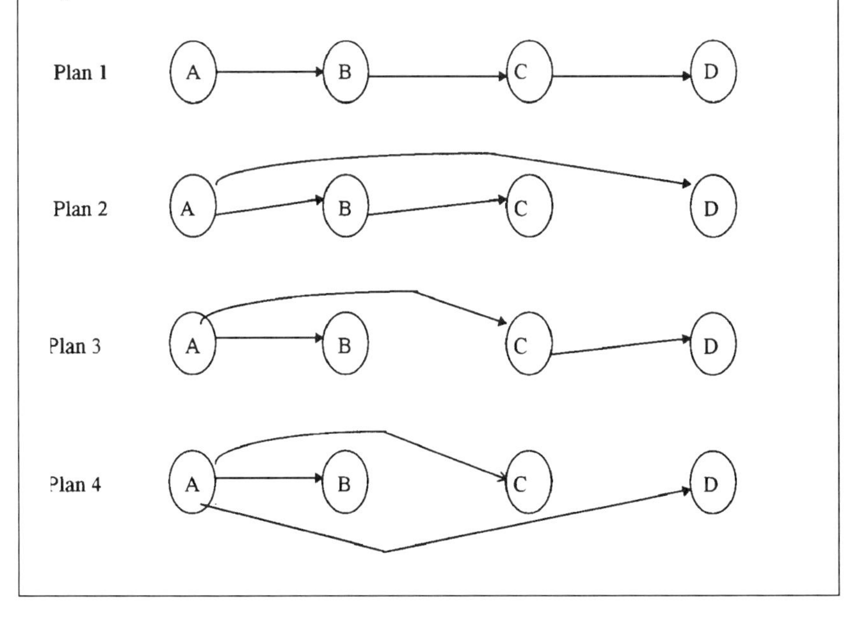

improve the existing blocking plans using optimization-based methods (Newton and Pam, 1996). The objective of the blocking problem is to minimize the total operating costs of delivering all traffic on the network while satisfying the resource and capacity constraints at the yards and the priority constraints for shipments. The operating cost covers both car-mile and car-hour costs, with a certain factor to convert these two types of cost. Yard capacity constraints are the physical resources available at the yards and are expressed in terms of the number of blocks that can be built and the number of cars that can be handled at each yard. Priorities for shipments are set by limiting the maximum number of reclassifications; for instance, automobile shipments may not exceed two reclassifications because of their time sensitivity whereas other merchandise may not exceed four reclassifications.

The notation used is as follows:

$G(N, A)$ is a graph with node set N and candidate block (arc) set A;
K is the set of all commodities;
v^k is the volume of commodity k (in consistent units);
δ_a^q is the incidence indicator that equals 1 if arc a is on path q and 0 otherwise;
ξ_i^a is the incidence indicator that equals 1 if i is the origin of a and 0 otherwise;
$Q(k)$ is the set of all legal paths for commodity k;
$E(k)$ is the maximum number of extra handling performed on commodity k;
$c_a \geq 0$ is the per unit cost of flow on arc a (assumed to be equal for all commodities);
u_a is the capacity of arc a;
$B(i)$ is the number of blocks which may originate at node i;
$V(i)$ is the volume which may be classified at node i for all commodities;
C_q^k is the path cost for flowing one unit of commodity k on path q.

The decision variables are as follows:

f_q^k is the proportion of commodity k on path q, $\forall\, q \in Q(k), k \in K$;

$$y_a = \begin{cases} 1 & \text{if block } a \text{ is included,} \\ 0 & \text{otherwise.} \end{cases}$$

The model is formulated as follows:

$$\sum_{k \in K} \sum_{q \in Q(k)} C_q^k v^k f_q^k \qquad (8.23)$$

subject to

$$\sum_{k \in K} \sum_{q \in Q(k)} v^k f_q^k \delta_a^q - u_a y_a \leq 0, \forall\, a \in A$$

$$\sum_{q \in Q(k)} f_q^k = 1, \forall\, k \in K$$

$$\sum_{a \in A} y_a \xi_i^a \leq B(i), \forall\, i \in N$$

$$\sum_{k \in K} \sum_{q \in Q(k)} \sum_{a \in A} v^k f_q^k \delta_a^q \xi_i^a \leq V(i), \forall\, i \in N$$

$$f_q^k \geq 0, \forall\, q \in Q(k), k \in K$$

$$y_a \in \{0, 1\}, \forall\, a \in A$$

where

$$C_q^k = v^k \sum_{a \in A} c_a \delta_a^q. \tag{8.24}$$

Commodity is defined as an origin–destination traffic pair. The objective function [equation (8.23)] minimizes total handling or classification cost. The number of handling for each commodity k on path q is obtained from equation (8.24). The first constraint is the bundle constraint relating blocks and path variables, which requires that no path can be used until blocks on the path are selected. The second constraint is the convexity constraint for each commodity. The third constraint represents the constraint on the number of blocks that can be built at each yard. The maximum number of cars that can be reclassified at each yard is represented by the fourth constraint. The last two constraints ensure non-negativity and integrality of the decision variables.

In the blocking problem, there are two sets of decision variables – block variable y_a and flow variable f_q^k. A feasible set of blocks will cover at least the path connecting the OD pair for each commodity whereas a feasible set of path flow variables will send all traffic from its origin to destination. Also, note that only the first constraint ties these two sets of variables together. By relaxing this constraint we can decompose the blocking problem into two subproblems, each of which involves only one set of decision variables. The two subproblems may be iteratively solved by using the concept of Lagrangian relaxation (Barnhart and Jin, 1997).

8.6 SUMMARY

Transportation decisions account for a large percentage of key logistics decisions in terms of both number and impact on costs. The first step in making good transportation decisions is to understand the costs involved in transportation. The second step is to bring in the other aspects of business such as inventory planning and warehouse location that are closely tied to transportation. In this chapter some of the key transportation decisions have been presented in a sequential manner. The primary reasons for the sequential treatment are tractability in modeling and solving for these decisions and simplicity in terms of understanding the process and solution. Some good start-up material on transportation has been provided in this chapter. The most common transportation decisions such as mode selection, route selection, scheduling, consolidation and fleet sizing are presented. In some cases, no analytical models have been included because of their complexity, which may be beyond the scope of this book. Some real-life examples of transportation models have also been included. There are a number of other real-life examples that are available in the literature. The quantitative modeling of transportation decisions is probably the most researched and published area in operations research. The importance of transportation decisions is also substantiated by the fact that almost all major transportation providers have full-time operations research staff to help make transportation decisions analytically.

PROBLEMS

8.1 Discuss a methodology for allocating fuel costs to a railcar traveling from city A to city B. Assume that the car travels through city C and changes train at that city.

8.2 Suggest an approach to estimate the operating cost of an aircraft flying from airport A to airport B.

8.3 The rate per hundredweight for a class 100 item to be transported 700 miles is $26.34, and the minimum charge is $40.32. Determine the transportation charges for an item rated at class 60 with a shipping weight of 100 lb and a travel distance of 700 miles.

8.4 The costs of travel between home base and three other cities that need to be visited by a traveling salesperson are given below. Determine the best sequence of visits that will minimize the total cost. The salesperson must visit all cities and each city must be visited only once.

From	*To*			
	home	*1*	*2*	*3*
Home	0	12	21	35
1	12	0	28	17
2	21	28	0	18
3	35	17	18	0

8.5 Determine the shortest path to go from city F to city C for the road network given in the figure in Example 8.5. The values along the links are the distances between the nodes at each end.

8.6 BCG America Inc. imports toys from the far east through its two major west coast ports, Los Angeles and San Francisco. The toys are shipped to the three stores in the east coast as soon as they are received, on a daily basis. The company also has two intermediate warehouses, one at Chicago and the other at Dallas. The daily import volumes at Los Angeles and San Francisco are 200 and 250 cases, respectively. The daily requirements at the three stores are 100, 150 and 200 respectively. The unit cost of shipping a case is given below. Determine the best routing for shipping toy cases from ports to stores.

From	*To*						
	Los Angeles	*San Francisco*	*Chicago*	*Dallas*	*Store 1*	*Store 2*	*Store 3*
Los Angeles	0	13	4	6	12	14	9
San Francisco	13	0	7	6	13	12	7
Chicago	4	7	0	3	8	8	12
Dallas	6	6	3	0	7	8	9
Store 1	12	13	8	7	0	17	11
Store 2	14	12	8	8	17	0	6
Store 3	9	7	12	9	11	6	0

8.7 Extend the fleet assignment model presented in section 8.4.3 to address multiple periods or days.

8.8 Formulate a mathematical model for locating hubs allowing only single allocation.

8.9 A transportation network has four cities on the east end and three cities on the west end. The company plans to build one or two hubs at the center of the network. With bidirectional travel determine the number of equipment that can be saved.

8.10 Discuss the similarities and differences between the fleet sizing problem and the automated guided vehicle capacity planning problem presented in Chapter 7, section 7.4.3.

REFERENCES

Ballou, R.H. (1992) *Business Logistics Management*, Prentice-Hall, Englewood Cliffs, NJ.

Barnhart, C. and Jin, H. (1997) A Lagrangian based procedure for railroad blocking problem. Research report, Massachusetts Institute of Technology, Cambridge, MA.

Campbell, J.F. (1994) Integer programming formulations of discrete hub location problems. *European Journal of Operations Research*, **72**, 387–405.
Fisher, M.L. (1981) The Lagrangian relaxation method for solving integer programming problems. *Management Science*, **27**, 1–18.
Hay, W.W. (1989) *An Introduction to Transportation Engineering*, Robert E. Krieger, Publishing Company, Melbourne, FL.
Jara Diaz, S.R. (1982) The estimation of transport cost functions: a methodological review. *Transport Reviews*, **2**, 257–78.
Kasilingam, R.G. and Hendricks, G.L. (1993) Cargo revenue management at American Airlines. *AGIFORS Cargo Study Group Meeting*, Rome, Italy; copy available from author.
Kasilingam, R.G. and Ong, C-H. (1995) Trailer requirements planning system for a major retailer. Technical report, Department of Industrial Engineering, University of Arkansas, Fayetteville, AR.
Kuhn, H.W. (1955) The Hungarian method for the assignment problem. *Naval Research Logistics Quarterly*, **2**, 83–97.
Muller, G. (1989) *Intermodal Freight Transportation*, 2nd edition, ENO Foundation for Transportation.
Newton, H.N. and Vance, P.H. (1996) *The Railroad Blocking Problem*, Technical Report, Auburn University, Auburn, AL.
Oakley, S.J., Kasilingam, R.G. and Hendricks, G.L. (1992) Air cargo overbooking under stochastic capacity. *Optimization Days 1992*, Montreal, Canada; copy available from author.
O'Kelly, M.E. (1986) The location of interacting hub facilities. *Transportation Science*, **20**, 92–106.
Reddy, V.R. and Kasilingam, R.G. (1995) Intermodal transportation considering transfer costs. *Proceedings of the 1995 Global Trends Conference of the Academy of Business Administration*, Aruba; copy available from author.
Singh, N., Aneja, Y.P. and Rana, S.P. (1990) Multi-objective modeling and analysis of process planning in a manufacturing system. *International Journal of System Science*, **24**, 621–30.
Wyatt, J.K. (1961) *Optimal Fleet Size*. Operational Research Quarterly, **186**.
Ziarati, K., Soumis, F., Desrosiers, S. *et al.* (1997) Locomotive assignments with heterogeneous consists at CN North America. *European Journal of Operational Research*, 281–92.
Zimmerman, H.J. (1978) Fuzzy programming and linear programming with several objective functions. *Fuzzy Sets and Systems*, **1**, 45–55.

FURTHER READING

Bott, K. and Ballou, R.H. (1986) Research perspectives in vehicle routing and scheduling. *Transportation Research A*, **3**, 239–43.
Council of Logistics Management (1994) *Bibliography of Logistics Training Aids*, 2803 Butterfield Road, Suite 380, Oak Brook, IL 60521.
Couillard, J. (1993) A decision support system for vehicle fleet planning. *Decision Support Systems*, **9**, 149–59.
Moore, J.L., Butler, D.P., Malstrom, E.M. and Kasilingam, R.G. (1996) Cost assessment of intermodal transportation linkages. *Proceedings of the AACE 40th Annual Conference*, Vancouver.
Winston, W.L. (1987) *Operations Research: Applications and Algorithms*, PWS-KENT, Boston, MA.

TRAINING AIDS: SOFTWARE AND VIDEOTAPES

Software

E-Z Rate Plus Software, Roadway Express Inc., tel. (216) 384-1717.
PC*Miler, ALK Associates Inc., tel. (609) 683-0220.
NUMERAX ONLINE Demo, Strategic Technologies Inc., tel. (908) 602-7268.

Videotapes

Manugistics Inc., tel. (301) 984-5000: *Manugistics Transportation Planning.*
Penn State Audio-Visual Services, tel. (800) 826-0312:

Business Logistics Management Series: Basic Transportation Rates;
Business Logistics Management Series: Link Analysis – An Overview;
Business Logistics Management Series: Special Transportation Rates and Services.

CHAPTER 9

Logistics performance metrics

9.1 INTRODUCTION

Logistics metrics are quantitative measurements that track certain processes within the logistics framework. The best design for a logistic system or component(s) of a logistics system truly depends upon the metric(s) used for measuring the performance. A system that measures up very high in one metric may not measure very well in some other criteria. The objective, however, is to design a system that meets or exceeds the expectations in most of the selected metrics. Logistics metrics vary based upon the boundary of the system (the various functional areas included such as production, distribution, inbound transportation, storage, vendor selection etc.), the functional requirements of the system and the different areas and the ability to define and measure them quantitatively. Hence the first step in designing the metrics is to define the system that needs to be measured and its components. The second step is to determine the functional requirements or expectations of the system. The third step is to identify metrics that can quantitatively measure the functional requirements. It is also important to understand the relationship between metrics. One or more metrics may drive the performance of another metric. For instance, in the case of railroads, customer service in terms of the percentage of on-time delivery of shipments depends upon the on-time arrival and departure of trains and terminal dwell time for cars (time spent at a terminal).

Ability to measure is absolutely essential to design and/or improve a logistics system or its components. Industries constantly strive to improve in several different areas in order to maintain or increase market share and profits. Hence the development of appropriate metrics is very critical to the success of any organization. Successful performance metrics are usually clear, simple and easy to understand. Metrics should reflect the important working dimensions of business operations. They should include the critical success factors for all levels of the business in terms both of financial and of non-financial measures.

Table 9.1 Results of a logistics metrics survey (Byers and Cole, 1996)

Logistics metric	*Ranking*[a]	
	Transportation	*Distribution*
On-time delivery	6.83	6.67
Complete and damage-free delivery	6.83	6.33
Accurate freight bills	6.17	5.8
Timely response to inquiry and claims	6	5.8
Order cycle time	5.5	6.6
Average transit time	6.4	5
Percentage backorders	4	6.14
Order accuracy and completeness	6	6.57
Customer communication	6	6.33
Service level	6.5	6.14
Inventory accuracy	5	6.5
Forecasting accuracy	5	5.75
Order selection	5	6
Order administration	4	5.67
Order shipping	4	6.2
Percentage loss or damage in storage	4	6.17
Throughput dollars per total inventory	0	6.4
Order throughput cycle time	6	6.8
Value of inventory adjustments	0	6.33
Replenishment cycle time	5	6
Percentage transaction processing errors	0	5.5
Number of orders shipped per year	1	5.5
Total logistics costs as a percentage of sales	6	6.25
Net profit as a percentage of sales	6	6.8

[a] Metrics were ranked by managers in the transportation and distribution functions, respectively. They were ranked on a scale of 1 to 7, with 1 being 'not important' and 7 'very important'.

Table 9.1 summarizes the results from a recent survey of metrics used by major US companies in the transportation and distribution functions (Byers and Cole, 1996). The results are based on the responses from people in transportation and distribution. The transportation column shows the average ranking (on a scale of 1 to 7, with 1 indicating 'not important' and 7 being 'very important') for the metrics included in the survey based on the responses from managers in the transportation function. The distribution column shows the average ranking for the same metrics based on the responses from managers in the distribution function.

The above survey illustrates the fact that the importance of the metrics vary depending upon the functional areas. For instance, the transportation managers ranked throughput dollars per total inventory and percentage transaction processing errors as 'not important'. Also, the metrics that were ranked the highest by the transportation managers – on-time delivery and complete and damage-free delivery – were not ranked at the top by the distribution managers.

9.2 CLASSIFICATION OF LOGISTICS METRICS

Logistics metrics are used to measure the performance of various logistics functions that are both internal and external to an organization. Metrics generally focus on aspects of time, quality, availability, cost, profit and reliability. Figure 9.1 presents a high-level classification of logistics metrics.

Internal metrics measure the performance of the system or the internal components of the logistics system such as production plant, warehouses and transportation equipment. Some of the examples of internal metrics are machine utilization, warehouse throughput and truck utilization. External metrics are measures that reflect the expectations of the organization by entities that are not part of the organization, such as customers, stockmarkets, government and third-party agencies. Examples of external metrics include response time to fix or replace defectives, the frequency of delivery and pick-up and percentage delivered on time. External metrics are generally defined at the appropriate level, depending upon the external parties involved. For a customer buying one of the company's services, the service will be measured on how well (time, cost and quality) it is provided. For a bank or investment company service will be focused on the system-level performance. It is to be noted that internal metrics drive the external metrics. The percentage of on-time deliveries of shipments by a transportation carrier is an external metric. However, it is driven by the internal metrics such as on time departure of delivery trucks, reliability of trucks, percentage of time on road etc. External metrics are typically financially and service orientated.

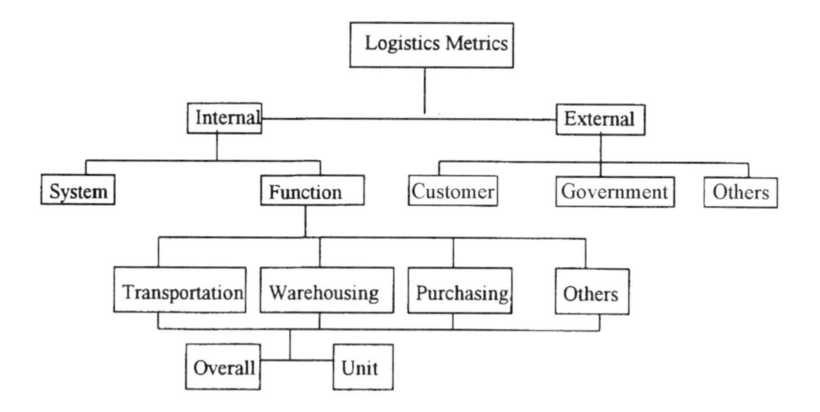

Figure 9.1 Classification of logistics metrics. Metrics may be financial or non-financial. Examples of financial metrics are cost and revenue; examples of non-financial metrics are service and productivity.

Metrics may be classified as strategic and operational depending upon the focus and level of detail. Strategic metrics focus on overall system-level performance whereas operational metrics focus on unit-level or machine-level performance. In both cases, the metrics used may be identical. When measuring equipment utilization, a system-level metric may focus on the utilization of all milling machines whereas a unit-level metric may focus on a particular milling machine individually. All metrics can be classified as financial or non-financial. Financial metrics may be for costs or revenue or profit, and non-financial metrics may be related to service and productivity or utilization. Commonly used financial metrics are unit transportation costs, operating ratio (expenses/net sales) and inventory turnover ratio. Typical non-financial metrics are machine utilization, percentage rejects and on-time arrivals. The following section provides a discussion of different types of logistics metrics.

9.3 LOGISTICS METRICS

Definitions and examples of a large number of metrics that are commonly used or applicable to the logistics system and the various functional areas within the system are presented in this section.

9.3.1 System-level internal metrics

System-level internal metrics focus on the financial or non-financial aspects of a company's performance. The following financial metrics will be discussed briefly: operating ratio, net profit ratio and inventory turnover ratio (Riggs, 1982). Asset utilization, total cycle time, system uptime, percentage defectives and percentage of demand satisfied and safety will be covered under the heading of non-financial metrics.

Financial metrics

Operating ratio

The operating ratio shows the percentage of dollars received from total revenue to meet the operating cost of running the business:

$$\text{operating ratio} = \frac{\text{total operating expenses}}{\text{total revenue}}$$

A lower ratio is preferred for obvious reasons.

Net profit ratio

The net profit ratio compares the profit after taxes with sales or total assets. With sales in the denominator this ratio indicates the level of sales required to make a certain amount of profit. When the denominator is the total assets the ratio is a measure of return on investment:

$$\text{net profit ratio} = \frac{\text{profit after taxes}}{\text{total sales}}$$

$$\text{net profit ratio} = \frac{\text{profit after taxes}}{\text{total assets}}$$

Inventory turnover ratio

The inventory turnover ratio is useful to evaluate the speed of movement of goods through a company. This ratio determines the number of times inventory is turned over during the year:

$$\text{inventory turnover ratio} = \frac{\text{cost of goods sold}}{\text{average inventory investment}}$$

Generally, a higher inventory turnover ratio is better.

Non-financial metrics

Asset utilization

Asset utilization is the amount of time for which a company's assets are working productively to generate the desired output:

$$\text{asset utilization} = \frac{\text{actual time working}}{\text{total hours in a period}}$$

Obviously, higher asset utilization is preferred. When the utilization is low, it may be a symptom of a host of other problems such as unreliable equipment, material shortage, power outage, etc.

Total cycle time

Total cycle time is the time from when an order is placed to the time when the shipment is received by the customer. It is a total of several different time elements:

$$\begin{aligned}\text{total cycle time} = \max\{&(\text{order processing time} \\ &+ \text{manufacturing lead time} + \text{transportation time}), \\ &(\text{order processing} + \text{delivery time from warehouse})\}\end{aligned}$$

It is desirable to reduce the cycle time in order to provide a better service as well as to improve asset utilization.

System uptime

System uptime shows the percentage of time the system is available for producing or delivering the required output or service:

$$\text{system uptime} = \frac{\text{time system is available}}{\text{total hours in a period}}$$

Percentage defective

The percentage defective indicates the quality of the production and inspection process of a company and is measured based on the number of defectives shipped versus the total number shipped to various customers:

$$\text{percentage defective} = \frac{\text{total defectives shipped}}{\text{total number of items shipped}}$$

This is an indicator of the output reliability of the system as well as a satisfaction measure for the customers.

Percentage of demand met

The percentage of demand met is an indicator of the operational capabilities of a company:

$$\text{percentage of demand met} = \frac{\text{orders fulfilled}}{\text{total demand}}$$

There may be several reasons for not being able to meet demand. Poor forecasting, insufficient capacity, low inventory levels, poor quality and unreliable production and distribution system are some of the reasons.

Safety

Safety may be measured in terms of the number of reportable injuries, accidents or fatalities as well as the number of hours lost because of them. Lack of safety costs everyone. Unsafe acts lead to damage to people, products and property, causing idle time of workers and facility and resulting in lower productivity and morale and higher costs. There are two popular metrics that are commonly used by most companies. The metrics may be used at the company level or at the functional or department level. The first is known as the frequency rate, which reflects the number of disabling injuries per million man-hours:

$$\text{frequency rate} = \frac{\text{number of lost time injuries}}{\text{number of man-hours worked}} \times 1\,000\,000$$

The second is known as the severity rate and represents the number of lost man-days per million man-hours worked:

$$\text{severity rate} = \frac{\text{number of lost man-days}}{\text{number of man-hours worked}} \times 1\,000\,000$$

The frequency rate assumes equal importance to all injuries whereas the severity rate takes into consideration the seriousness of the injuries in terms of lost man-days.

9.3.2 Function-level internal metrics

In this section, discussion is given on a few select metrics that are related to some of the important functions in logistics – transportation, warehousing, inventory control, production, maintenance and vendor selection. The metrics presented here, in most cases, are suitable for overall performance measurement as well as unit-level performance measurement. For instance, locomotive utilization as a performance metric can be applied to the entire fleet of locomotives, or to locomotives of a particular type, or to a specific locomotive. The financial and non-financial metrics in each of the functional areas are presented.

Vendor selection

The metrics discussed under this category are very useful in constructing vendor profiles. The objective of these metrics is to evaluate the available vendors in terms of quality, cost and delivery reliability.

Percentage of good parts

The percentage of good parts is a measure of the reliability of the supply in terms of the quality of the delivered parts:

$$\text{percentage of good parts} = \frac{\text{total supply} - \text{defectives}}{\text{total supply quantity}}$$

Ideally, one would like to have this number equal to 100%. Otherwise, it leads to shortage of parts and additional costs for rework or returning them to the vendor.

Unit cost

Unit cost typically includes the manufacturing, packaging, handling, storage and transportation costs. All others being equal, selection of vendors will be clearly based on unit costs, assuming no quantity discounts. When quantity discounts exist, selection will be based on total delivered cost:

$$\begin{aligned}\text{unit cost} = {} & \text{unit overhead cost} + \text{unit costs of (manufacturing} \\ & + \text{packaging} + \text{handling} + \text{storage} + \text{transportation)}\end{aligned}$$

Delivery reliability

Delivery reliability indicates the variability in delivering a buyer's order. There are two things that are important for a buyer. The first is a shorter promised delivery time. The second is having zero variability around it. Most customers would prefer a lower variability and longer delivery time to a highly variable shorter delivery time. A simple formula to compute variability is as follows:

$$\text{delivery reliability} = \frac{\text{maximum delivery time} - \text{minimum delivery time}}{\text{average delivery time}}$$

Other formulas based on standard deviation provide a much more accurate measure of delivery reliability.

Complete shipments

The complete shipments metric measures the percentage of orders that are delivered in full. An order is not considered delivered in full if it does not have the right items in the right quantities:

$$\text{complete shipments} = \frac{\text{number of orders delivered in full}}{\text{total orders in a period}}$$

Transportation

Transit time

Transit time is a measure of the time to move a shipment from origin to destination. The origin or destination may be any of the locations within the logistics system such as the customer, store, warehouse or plant. For the transportation industry this is commonly known as the dock-to-dock time:

$$\text{transit time} = \text{travel time} + \text{waiting time at terminals or docks} + \text{transfer time} + \text{handling time}$$

Transfer time is the time to move the shipment from one mode to another or from one vehicle to another vehicle (truck, train, aircraft or ship). Handling time refers to the time to load or unload the shipment at the origin and destination.

Transit time variability

The transit time variability measure captures the variation in transit time. It is an indicator of the reliability of the transportation function. As discussed before, it may be measured by using the range or standard deviation of transit times:

$$\text{transit time variability} = \frac{\text{maximum transit time} - \text{minimum transit time}}{\text{average transit time}}$$

Again, most customers prefer less variability in transit time since it helps them to plan their work more efficiently.

Transportation cost per unit

Transportation cost per unit represents the cost for transporting one unit of the product or part. Other similar measures include transportation cost per ton, transportation cost per trailer etc. The cost of transportation includes the fixed and variable components of the transportation function. Several different methods are available to allocate the fixed costs and to account for backhauls. Some of the approaches were presented in Chapter 8.

Damage-free shipments

The damage-free shipments metric is an indication of the quality of the transportation service. Obviously, lower damage benefits both the customer and the service provider. Lost shipments are considered as fully damaged and non-repairable:

$$\text{damage-free shipments} = \frac{\text{number of damage-free shipments}}{\text{total number of shipments}}$$

Perfect shipments

This is an overall quality measure of the transportation function. It is based on the number of shipments that arrived at the final destination on time, complete, damage-free and with complete documents:

$$\text{perfect shipments} = \frac{\text{number of perfect shipments}}{\text{total number of shipments}}$$

Equipment utilization

Most transportation equipment is very expensive. Equipment includes the motive power and the trailers or cars. It is very important to maximize the utilization of the equipment. Higher utilization means a lower cost of moving goods and/or a faster turnaround time of equipment.

Equipment utilization may be measured in several ways. Trailer or car utilization is calculated as:

$$\text{trailer or car utilization} = \frac{\text{loaded travel time}}{\text{total travel time}}$$

Total travel time includes waiting time, loading and unloading time and actual travel time. Alternatively, one may use:

$$\text{motive power utilization} = \frac{\text{active pulling time}}{\text{total time}}$$

Active pulling time refers to the time during which the tractor or locomotive is pulling trailers or cars. Total time includes active pulling time, maintenance time, deadheading time and other waiting time

Facility utilization

Facility utilization measures the efficiency of the facility that handles the equipment (cars or trailers and aircraft, locomotives, tractors and tug boats). It may be an airport, railroad terminal, cross docking facility or a port. This metric measures the time between arriving at and departing from a facility, which is commonly known as ground time or dwell time. Most companies monitor the actual dwell time against the planned dwell time:

$$\text{car or trailer dwell time} = \text{departure time} - \text{arrival time}$$

$$\text{locomotive dwell time} = \text{departure time} - \text{arrival time}$$

On-time arrival and departure

This is one of the most popular metrics used by transportation companies. The definition of on-time may vary from industry to industry and within an industry from company to company. An airline may measure within a 5 min window whereas railroads may measure within 4 hours:

$$\text{on-time-arrival or departure} = \frac{\text{number of on-time arrivals or departures}}{\text{total number of arrivals or departures}}$$

Inventory control

In the following, measures related to cost and service are presented (Tersine, 1994). Cost-related metrics are used as indicators of the financial performance of the inventory control function. Service metrics measure the customer focus of the function. Customer focus is measured in terms of time and availability.

Inventory cost per unit

Inventory cost per unit includes the cost of carrying a unit inventory and storage costs. Storage costs include rent and taxes and other costs that are required to protect the inventory such as heating and lighting, insurance and security. It also includes the cost of obsolescence.

Aggregate inventory value

Aggregate inventory value represents the total value of inventory at cost. This is a measure of the amount of inventory investment. Most companies set an upper limit on this amount. This is a simple and easy-to-use metric but fails to capture the dynamic aspects of inventory.

Inventory turnover

Inventory turnover has been discussed in section 9.3.1 under the subheading of financial metrics.

Demand not met

Demand not met reflects the performance of inventory control from a customer perspective in terms of availability:

$$\text{unsatisfied demand} = \frac{\text{demand not met in time}}{\text{total demand}}$$

Unsatisfied demand leads to dissatisfied customers, causing market erosion. Sometimes customers may be willing to wait to receive the goods from a supplier because of lack of competition or the superior quality of the products. In this case, additional expenses are incurred to expedite the demand fulfilment process.

Warehousing

The main function of a warehouse is to facilitate the storage and retrieval of items as and when needed. The four major metrics of warehouse performance are related to cost, utilization, time and quality.

Time

The time to retrieve an item from storage directly translates to customer service. Order picking time is a measure of how much time it takes to pick all the items listed in a customer's order. Companies adopt several different approaches to determine the storage locations of the different items based on their value, frequency of use and volume of usage:

$$\begin{aligned}\text{order picking time} = {} & \text{order processing time} + \text{travel time to first location} \\ & + \text{interlocation travel time} \\ & + \text{travel time from last location} + \text{pick-up time} \\ & + \text{interference or waiting time}\end{aligned}$$

Order processing includes determining the locations of the items and planning the routing sequence to pick up the items. Interference or waiting time relates to the time spent in waiting for material handling equipment, stops during actual movement as a result of congestion and so on.

Equipment utilization

This metric is very similar to the equipment utilization metric discussed under transportation above but applies to the material handling equipment.

Warehouse throughput

Warehouse throughput can be defined as the average number of loads per hour a storage system can receive or place in storage and retrieve from storage (Ballou, 1992). The system has to be designed for the maximum expected throughput that will occur during a day. Order picking time combined with utilization of the material handling system impacts warehouse throughput.

Warehouse operating cost per unit
Warehouse operating cost per unit is a measure of the cost-effectiveness of operating a warehouse and is impacted by warehouse size and throughput. Fixed cost components increase with warehouse size. Fixed cost components include building, equipment and fixed payroll. Variable cost components increase with the number of loads processed; however, the combined unit cost goes down. Variable components include contract manpower, variable utilities, fuel etc. Overhead expenses include heating, lighting, insurance, taxes etc.

Production and manufacturing

Metrics in this function measure the utilization of the production equipment, quality of product from the production line and the associated cost.

Production rate
The production rate is the number of units that can be produced in a shift or day or during a specified period. The production rate determines the manufacturing cycle time (Groover, 1987). It depends upon the design capacity and availability of the production equipment and on the quality of the production process:

$$\begin{aligned}\text{production rate} = {} & (\text{design production rate}) \times (\text{percentage defective}) \\ & \times (\text{percentage availability})\end{aligned}$$

Production time per unit is the reciprocal of production rate.

Production cost per unit
The production cost per unit is the total cost of labor, material, tools, equipment and overheads to manufacture one unit of product. Factored into this are actual production cost, set-up cost and tool and equipment changeover costs. Unit production is influenced by the number of good quality parts produced.

Defectives
Some of the defects may be fixed. A defective containing repairable defects may be turned into a good finished product without seriously impacting the production rate. Defectives with defects that cannot be fixed have to be scrapped, bringing down the production rate.

Defectives increase the unit cost of production because of the costs incurred to repair the defects and the cost of lost labor and material on the rejects:

$$\text{percentage rejects} = \frac{\text{number of units that cannot be repaired}}{\text{total number of units produced}}$$

Machine or facility utilization

The machine or facility utilization metric is very similar to the equipment utilization metric discussed above under the heading of transportation but applies to the production equipment and facilities.

Maintenance

Maintenance is a necessary evil to operate any system efficiently. Measures of maintenance focus on the uptime of equipment, time to repair when equipment is broken and cost of maintenance.

Logistics delay time

Logistics delay time is the maintenance downtime as a result of waiting for a spare part to become available, waiting for transportation, waiting for the use of a facility to become available, waiting for the technician to arrive and so on (Blanchard, 1992). This does not include the actual active maintenance time:

$$\text{logistics delay time} = \text{sum of all waiting times arising from the non-availability of resource(s) on time}$$

Mean time to repair

The mean time to repair, $\overline{t^{r}}$, is the average active maintenance time to perform corrective maintenance. This depends upon the type of failure and the frequency of failures. It is calculated as follows:

$$\overline{t^{r}} = \frac{\sum f_j t_j}{\sum f_j}$$

where f_j is the frequency of failure type j and t_j is the average time to repair failure type j.

Mean time between failures

The mean time between failures, $\overline{t^{f}}$, is the average time between any two failures. The time between two successive failures is the difference between the time the previous failure was fixed and the time at which the current failure occurred. It can also be defined as the average of all uninterrupted system uptimes:

$$\overline{t^{\mathrm{f}}} = \frac{1}{n}\sum t_j^{\mathrm{u}}$$

where t_j^{u} is the uptime between failure $j-1$ and j, and n is the number of failures.

Maintenance cost per unit produced
The maintenance cost per unit produced includes both preventive and corrective steps taken to keep production, transportation and handling equipment in good condition in order to manufacture quality products and to transport products safely and without damage. Maintenance costs include labor, supplies, spares and other overheads.

9.3.3 External metrics

External metrics provide an outsider's view of the financial and non-financial aspects of a company's performance. The outsiders are usually customers, government and investors.

Customer service

Customer service essentially has three dimensions: quality, time and cost. Quality may be measured or expressed in several ways depending upon the type of industry. For a manufacturing industry, it may be expressed in terms of the number of good parts or products; for the transportation industry it may be the number of shipments delivered on time. Time-based metrics capture the waiting time of the customer. In a manufacturing industry it is the lead time (time between order and receipt) to manufacture and deliver. In the hospital industry it is the time between arrival and departure after treatment. In the retail industry it may be the availability of a product within a certain time. Cost-based metrics essentially indicate the unit cost of manufacturing or providing a service. The list of customer service metrics is truly endless because of the different types of industries. Some of the function-level or system-level internal metrics may also serve as useful customer service metrics. Examples include on-time arrival and departure, manufacturing lead time, transportation cost and manufacturing cost. Only a couple of customer service metrics will be discussed in this section.

Service reliability
Most transportation companies use this metric to measure their customer service performance. It is measured by the number of shipments that are delivered to the customer within certain time units of the promised delivery time. The time unit may be days, hours or

minutes. For instance, in railroads this may measured by ±4 h within the promised delivery time:

$$\text{service reliability} = \frac{\text{number of shipments within } \pm 4\,\text{h of promised delivery time}}{\text{total number of shipments}}$$

Customer complaints

Customer complaints may arise from a variety of reasons. Late arrival, late departure, not delivering on time, a lost or damaged shipment, inaccurate order and poor quality parts are some of the common reasons. Most companies aggregate all the complaints:

$$\text{customer complaints} = \text{number of complaints during a period}$$

Some companies separate them based on the functional areas.

Government perspective

The measures used by the state government may not be the same as those used by the federal government. County office metrics may differ from city-level metrics. Governments tend to measure industry performance based on the number of new jobs created, revenue from taxes, new infrastructure development and the impact on other industries. Governments also evaluate the negative aspects such as pollution, congestion and crime. For instance, opening up an entertainment park creates employment, leads to the development of new roads and creates other supporting industries such as restaurants and hotels. The negative impact could be traffic congestion and parking problems.

Perspectives of investors

Investors and shareholders have an entirely different view of the performance of a company. It is almost always based on monetary terms. Two financial metrics that represent shareholders' interests are presented below (Riggs, 1982).

Earnings per share

The earnings per share indicates how well a company performs in terms of retained earnings (after tax operating income) with respect to the total number of shares of stock. A higher value means better performance:

$$\text{earnings per share} = \frac{\text{retained earnings}}{\text{number of shares}}$$

Financial leverage

Financial leverage is the ratio of total debts to total assets:

$$\text{financial leverage} = \frac{\text{total debts}}{\text{total assets}}$$

A company having $100 million in assets and $60 million in debts has a leverage factor of 0.6. Higher leverage means that more money is owed outside the firm, implying a higher rate of return on the amount owners have invested. However, it also increases the uncertainty about future returns to the shareholders since a large portion of the assets are financed by debts.

9.4 ISSUES IN LOGISTICS METRICS

In this section discussion is given on some of the issues involved in understanding and implementing logistics metrics. The following list covers most of them:

- basic criteria for metrics – simple and easy to understand in terms of measuring, calculating and interpreting;
- measurement units (scales of measures) and data availability;
- how often to measure and report and who gets the reports;
- reporting format – charts or graphs (target against actual) and trend analysis;
- modeling of the relationship between metrics and setting targets for related metrics;
- departments or entities involved.

9.4.1 Criteria for metrics

Metrics in general should be very simple and easy to understand. This requirement is even more important to logistics metrics because of the interdepartmental nature and the involvement of external agencies in various logistics functions. The objective of any metric is to measure the performance of a function or an entity, monitor the performance and take corrective actions when performance stays poor or improves at a slower rate. Also, it is equally important to understand the reasons for good or outstanding performance in order to retain it over an extended period of time. These objectives may only be achieved through the use of simple metrics. Ease of understanding is critical to sell the metric to all concerned. By definition all metrics must be focused and measurable. Measurability implies that it can be defined quantitatively and that data are or will be available to implement it. It is very important to minimize the number of metrics used – whenever possible, identical and similar metrics must be consolidated. The use of a large number of metrics reduces the effectiveness of the usage of all the metrics, in particular the key metrics. Metrics of a similar nature across different departments within an organization must be compatible. For example, if the efficiency of the transportation department is measured based on transportation cost per unit, then all other departments must also be measured in terms of unit costs.

9.4.2 Measurement units

Defining the units of measurement and the scales used for comparing the metrics over a period of time or from different departments is very crucial. Measurement units should bring out the variability in performance as well as be easy to manipulate. For instance, transit time in minutes may bring out the variability better than transit time in days; however, it may involve very large numbers to represent even the shortest trips. The scales used for representing the values of the metric in the form of a graph or chart should be meaningful. Whereas a scale of 1 to 5 may not exhibit any variation, a scale of 2 to 1 may overemphasize it. Let us say the travel times to go from point A to point B in two different periods are 70 min and 80 min, respectively. With a scale of 1 to 5, these values will be represented as 14 and 15, underemphasizing the variation; with a scale of 2 to 1 they will be 140 and 160, overemphasizing the variation. The availability of data or the ability to collect data at a certain level also dictates the units of metrics to be used.

9.4.3 Reporting of metrics

When metrics are implemented there are a few key areas that need to be addressed. The first is where do the inputs to the metric come from? How are they transmitted? Is the computation of the metric done manually or by computer? How frequently should the metric be reported – daily, weekly, monthly, quarterly or yearly? Who gets the reporting on the metrics – personnel at what level and from which departments? Should the distribution be done electronically? Is the reporting in the form of a table or graph? Must any comparative or trend information be provided – showing the utilization of a milling unit over a period of time or comparing it to the utilization during the same periods last year or against a target? The reporting frequency depends upon the level of detail at which a function is managed. High-level metrics may be reported more often than detailed ones. The cost of reporting should be weighed against the benefits of reporting. With the current pace of computer hardware and software development electronic distribution seems to be the obvious media of reporting. The

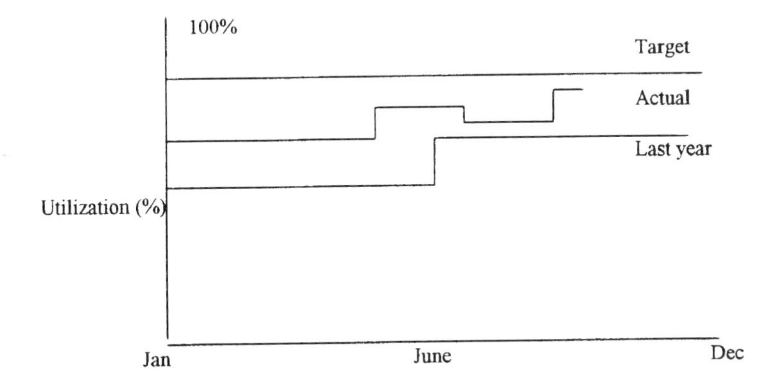

Figure 9.2 Performance monitoring of paint shop utilization.

reporting of certain metrics may have to be posted at certain key locations as information to all employees. It is always helpful to display performance metrics as a graph with comparative statistics. Figure 9.2 shows an example of reporting the utilization of a painting shop.

9.4.4 Understanding the drivers of metrics

The performance of an entity or a function is often driven by certain other factors that may be internal and/or external to it. It is important to understand the drivers of the performance of any metric. In real life it is always helpful to develop metrics that measure the performance of the drivers. This ensures consistency and validates the whole concept of measurement and monitoring. The development of analytical models that express the relationship between metrics is often a requirement to sustain or improve overall performance. These models will enable a company to answer two types of questions: (1) Given certain performance levels of the drivers, what will be the performance of the overall (driven) metric? (2) What should be the performance levels of the drivers in order to achieve a certain overall performance?

A simple example from railroad operations is used to illustrate this. The dock-to-dock service reliability or the percentage of shipments that are delivered within x hours of promised time is often used as one of the metrics to indicate the level of service provided to the customers. Other metrics may be dock-to-dock transit time, cost etc. The major drivers of service reliability are the performance of local trains at the origin and destination, terminal performance and the line of road performance. Railroads constantly try to improve the pick-up and delivery of cars by the locals at the origin and destination yards, the on-time departure and arrival of line of road trains at major terminals and the on-time handling of cars at the terminals. These are achieved through use of better operating plans, handling procedures, locomotive assignments and train schedules. A tool is needed to predict service reliability given the performance of locals, line of roads and terminals. Also, in order to achieve a target service reliability the railroad needs to know the breakeven performance levels of the locals, line of roads and terminals.

The generic model to predict service reliability may be expressed as follows:

service reliability $= f\{$origin local, destination local, terminal, line of road$\}$

The actual functional form of the model for a given railroad may be as follows:

$$\text{service reliability} = (P^{\text{OD}})^a T^b L^c$$

where, P^{OD} is the combined performance of origin and destination locals, T is the terminal performance and L is the line of road performance; a, b and c are the corresponding exponents. An inverse of the service prediction model may be used to compute the breakeven performance levels of the three components. Figure 9.3 shows an example of breakeven performances for a target service reliability of 80%.

Development of the inverse model for computing the breakeven performance levels of the drivers is often very difficult and complex. The primary reason is that the models to predict the performance of the overall performance are themselves often complex mathematical or statistical models. More and more companies have started using these models to set performance targets and to develop programs to achieve the targets.

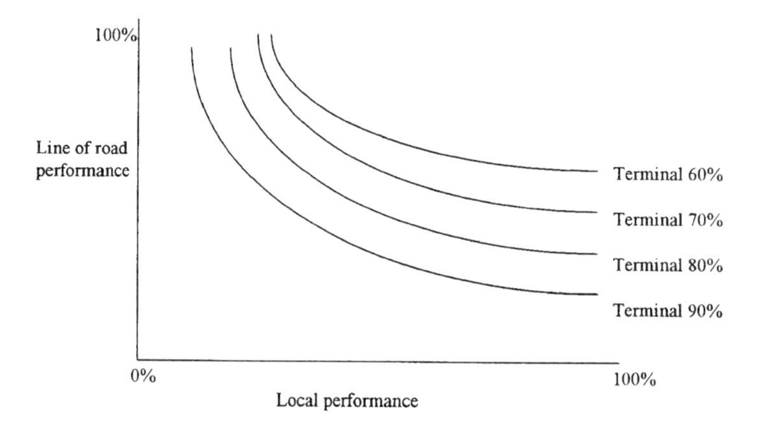

Figure 9.3 Isographs for 80% service reliability.

9.5 SUMMARY

In this chapter, attempt has been made to present an overview of the various types of metrics that are useful within the logistics and transportation framework. A classification of the logistics metrics has also been suggested. Almost all the metrics center around cost, around quality, service or reliability and around productivity. Some of the metrics are very fundamental and common to all the functions within logistics. Examples are unit cost and cycle time. Purchase cycle time includes the time between ordering and receiving the items; manufacturing cycle time is the time between receiving the order and completing the manufacturing of the product; transportation cycle time is the time to move a part or product from its origin to its destination. Similarly, one can have purchase cost, manufacturing cost, transportation cost, storage cost etc. The metrics discussed here are comprehensive but are by no means exhaustive. The objective of this chapter is only to provide guidelines for developing logistics metrics. The exact nature and definition of metrics will greatly depend upon the type of industry, the primary business of the company, the long-term and short-term goals of the company and so on.

PROBLEMS

9.1 Develop a list of possible metrics that may be used by a retail corporation.

9.2 Apply the classification scheme shown in Figure 9.1 to an automobile manufacturing company.

9.3 A production line is operated 60 hours a week. Its design capacity is 20 units per hour. On an average, the line is up 85% of the time and produces 3% defectives. Determine the production rate of the line.

9.4 The net sales of XYZ Corporation in 1996 was $240 000. The total expenses were $200 000 and the average inventory was worth $40 000. Compute the operating ratio and inventory turnover ratio.

9.5 A warehouse forklift had six failures in the first six months of its operation. The forklift operated for eight hours a day. The first failure occurred after 200 hours, the second after 280 hours, the third after 600 hours, the fourth after 800 hours, the fifth after 1000 hours and the last one after 1260 hours. It took nearly 10, 14, 8, 40, 20 and 50 hours, respectively, to fix the six failures. Compute the mean time to repair and the mean time between failures.

9.6 A city bus operating between downtown and the airport has a scheduled run time of 36 minutes. The actual run times for 6 trips on a given day are 34, 40, 42, 35, 39 and 41 minutes, respectively. What is the reliability of the service?

9.7 The arrival and departure times of a truck at a retail store for the past five deliveries are given below:

Arrival	0910	0900	0850	0905	0855
Departure	1021	1022	1010	1015	1005

Compute the average dwell time for the truck.

9.8 QuickChange is an automobile garage specializing in balancing, changing and rotating tires. Suggest five important metrics that will drive QuickChange toward capturing and retaining business.

REFERENCES

Ballou, R.H. (1992) *Business Logistics Management*, Prentice-Hall, Englewood Cliffs, NJ.

Blanchard, B.S. (1992) *Logistics Engineering and Management*, Prentice-Hall, Englewood Cliffs, NJ.

Byers, J. and Cole, M. (1996) Logistics metrics. Interim research report, Department of Industrial Engineering, University of Arkansas, Fayetteville, AR.

Groover, M.P. (1987) *Automation, Production Systems, and Computer-integrated Manufacturing*, Prentice-Hall, Englewood Cliffs, NJ.

Riggs, J.L. (1982) *Engineering Economics*, McGraw-Hill, New York.

Tersine, R.J. (1994) *Principles of Inventory and Materials Management*, Prentice-Hall, Englewood Cliffs, NJ.

FURTHER READING

Blumberg, D.F. (1994) Strategic benchmarking of service and logistics support operations. *Journal of Business Logistics*, 89–119.

Byrne, P.M. and Markham, W.J. (1991) *Improving Quality and Productivity in the Logistics Process*, Council of Logistics Management, 2803 Butterfield Road, Suite 380, Oak Brook, IL 60521.

Tompkins, J.A. and Harmelink, D. (Eds) (1994) The *Distribution Management Handbook*, McGraw-Hill, New York.

CHAPTER 10

Recent trends in logistics

10.1 INTRODUCTION

All the foregoing chapters presented some of the commonly known areas in transportation and logistics. This chapter presents some of the most recent topics or areas in logistics that have become very popular and have gained importance from the practitioner's point of view. The intent of this chapter is not to cover all of the latest concepts or trends in logistics but to highlight some of the important aspects of certain key topics. For an extensive and detailed discussion of recent topics, readers are referred to the appropriate references listed or the further reading section at the end of this chapter. The following five topics are presented in this chapter:

- third-party logistics – the reasons for outsourcing, a process for implementing third-party logistics, and some of the impediments to third-party logistics are addressed;
- benchmarking – different types of benchmarking and the process of benchmarking are presented;
- reverse logistics – the need for environmentally responsible logistics or reverse logistics, the options for implementing reverse logistics and the steps required to implement reverse logistics are discussed;
- global logistics – flows in global logistics and a generic model for the global supply chain are presented;
- virtual warehousing – an introduction to virtual warehousing is presented.

10.2 THIRD-PARTY LOGISTICS

A third-party logistics service is something more than subcontracting or outsourcing. Typically, subcontracting or outsourcing covers one product (or a family of products) or one function that is produced or provided by an outside vendor. Examples include automobile companies subcontracting the manufacture of airbags, or construction companies subcontracting roofing, or retail companies outsourcing the transportation function. Third-party logistics providers cut across multiple logistics functions and primarily coordinate all the logistics functions and sometimes act as a provider of one or more functions. The primary objectives of third-party logistics providers are to lower the total cost of logistics for the supplier and improve the service level to the customer. They act as a bridge or facilitator between the first party (supplier or producer) and the second party (buyer or customer).

10.2.1 Reasons for outsourcing

There are several reasons for the growth of third-party logistics over the past decade. The transportation and distribution departments of some of the major corporations have been downsizing in order to reduce operating costs. The most logical area to reduce costs is advisory functions such as operations research, followed by support functions such as transportation or warehousing. The area where companies want to strengthen by investing more is their core competency. Though it may sound like a fad it has been a reality at some of the major corporations. The other reason is from the customer side. Customers demand an exceptional service but are not willing to pay an extra-ordinary price for it. This requires the use of faster and frequent transportation services and flexibility in inventory levels. A third-party logistics provider will be in a position to consolidate business from several companies and offer frequent pick-ups and deliveries, whereas in-house transportation cannot.

Though downsizing and cost reduction have been cited by some companies as reasons for outsourcing, the decision to outsource should not be based on such considerations. The decision should be based on other important issues, as listed below (Canitz, 1996):

- the company's core business or competency may not be in logistics;
- sufficient resources, both capital and manpower, may not be available for the company to become a 'world-class' logistics operator;
- there is an urgency to implement a 'world-class' logistics operation or there is insufficient time to develop the required capabilities in-house;
- the company is venturing into a new business with totally different logistics requirements;
- merger or acquisition may make outsourcing logistics operations more attractive than to integrate logistics operations.

Armstrong (1996) presents a structured profile of third-party logistics providers. The major third-party logistics providers may be grouped into three categories: (1) distribution chain of high-value commodities with international orientation; (2) integrators of inbound logistics; and (3) dedicated contract carriers with service limited to North America.

Different providers have different capabilities in terms of freight bill payment, import and export, site selection, dedicated contract carriage of goods, management of the supply chain and international coverage. Similarly, most providers tend to be associated with certain types of commodities. The number of tractors in the dedicated contract fleet varies from 25 to 6000. The number of employees associated with logistics functions range from 10 to 7000. The net logistics revenue varies from $1 million to $860 million.

The third-party logistics market is a very dynamic one. This is because of a number of reasons. The primary reasons are changes in employee resources of providers, changes in expectations from the buyers of third-party services and changes in the offer of the providers to attract more customers and to retain or increase profit margins.

10.2.2 Process of implementing a third-party logistics service

The process of implementing a third-party logistics service should be rigorous and thorough in order to attain maximum benefits from it. An approach involving four phases is presented here. The first phase is concerned with defining the scope of the services to be outsourced and preparing a request-for-proposal (RFP) to be sent out to prospective providers. The second phase focuses on identifying candidate providers and evaluating them based on some criteria in order to select a provider. The third phase involves negotiating and finalizing the contract in terms of expectations, time period and price. The final phase is concerned with managing and monitoring the outsourcing process.

Phase 1: scoping of requirements

This is the first and foremost of the steps involved in selecting a good logistics provider. The buyer must identify the logistics functions that need to be outsourced. The coverage of the functions in terms of geographical area, customers and commodities also have to be well-defined, for instance the outsourcing of management of the warehousing function for international customers of computer printers. Now, how does one identify the functions and their coverage that are best suited for outsourcing? In general, an analysis of strengths and weaknesses of the buyers coupled with the requirements from the vendors will usually help in this process. Benchmarking against other similar companies may also help to identify functions to be outsourced. This analysis when combined with the availability or acquisition of resources results in the potential candidate functions to be outsourced. Then, the internal departments that are currently responsible for these functions have to be bought into the idea of outsourcing. Typically, people are resistant to change for a number of reasons: loss of control, job security, uncertainty in general and a 'why me?' attitude. The company may have to go through a change in management process to cross this stage.

Once this stage is crossed, the next step is to prepare an RFP. The RFP document clearly outlines the customer requirements for the functions to be outsourced. It also includes the expectations from the provider of the service. At this time metrics that will be used for monitoring performance should also be developed. An RFP should include general information about the company, information about customers and logistics requirements (storage, material handling and transportation). It should also contain specific sections for the potential providers to include organizational background, pricing data, capabilities and a list of their customers.

Phase 2: selection of service provider

This phase involves two stages. The first stage involves identifying potential providers for the functions to be outsourced. A list of potential service providers for a logistics function may be found from trade magazines, published surveys and web sites. The second stage involves selecting one or a few providers actually to contract out the service. This includes evaluation of proposals, site visits, obtaining references and then combining all these together analytically to short list the final candidates. The analytical results may then be combined with experiential knowledge and judgment to select the actual providers.

Phase 3: implementation and contracting and follow-up

This is a very crucial stage since it involves the actual contract. The contract should cover service requirements, price, payment schedule, duration and other specific exclusions and inclusions. The specific clauses may cover issues related to contract cancellation, price increases and sharing of cost savings. The type of metrics that will be used for measuring service and the frequency of measurement should be included. The type and extent of support that will be provided by the company to the provider and training requirements should also be stated clearly.

Phase 4: management and monitoring

This is a very critical step to continue to improve the relationship between the provider and the company as well as to measure the performance of the provider and the company in terms of the expectations set out in the contract. This may be done at formal meetings or through quarterly reports. The buyer of the third-party service should dedicate a team to work with the provider. This team will be the point of contact for handling and managing all the issues related to the contract or agreement. Issues of poor performance by the provider and inadequate support from the company should be reviewed periodically to identify the barriers and fix them. Failure to do this will lead to an unhealthy relationship and eventually may lead to the termination of the contract.

10.2.3 Impediments to third-party logistics growth

Like any business transaction, the business of third-party logistics involves a buyer and a seller. The seller is the third-party logistics company providing the required logistics services and the buyer is the client company using the services. The impediments stem from the expectations, perceptions and experiences of the buyer and seller. Hence the impediments to third-party logistics growth are broadly categorized into two areas (Maltz and Lieb, 1995): the limitations of the buyers and the limitations of the third-party logistics providers.

Limitations of the buyers

The first limitation is related to measurement. Companies outsource one or more of their logistics functions primarily to reduce logistics costs. It is usually simple and easy to quantify the expectation and the realization if all the logistics functions are outsourced. However, most companies outsource only a few functions. The problem with this is that it is hard to separate out and track the relevant logistics costs for the outsourced functions. To improve any process one must have certain metrics that are clear, simple and easy to use to measure success. The non-availability of metrics may be the result of a number of reasons: complexity of the function outsourced, non-existing current data on performance and lack of a suitable mechanism for data collection and reporting.

The second limitation is related to the selection of the function(s) to be outsourced. The functions outsourced may not always be the ones which need an outside vendor either from the point of view of cost cutting or service improvement. It may very well be based on which departments can be sold on the idea of outsourcing. An associated problem is the use of multiple providers for different functions. This leads to lack of coordination among the providers and no one being held accountable or responsible for the outcome. Also, it is hard to measure one provider in isolation of the other. The advantages of synergy are lost with this type of arrangement. The buyer ends up spending time and money managing and maintaining communication among the providers.

The third limitation is related to expertise. Not all third-party providers are experts in all logistics functions. Hence a thorough process must be established to select providers. In-house employees must be involved in the process since they know the company's products and its logistics requirements better than anyone else. Even after signing the contract with a third-party logistics provider, the company should maintain a group of in-house experts to monitor the performance of third-party providers and to assist them with any information requirements.

Limitations of the third-party logistics providers

The providers of third-party logistics services have to meet a number of challenges in order to survive through lower prices arising from competition and demand for continuous improvement in service from the buyer. The first limitation is competition from other third-party logistics providers and from the internal departments of the client. The fierce competition forces the provider to lower prices. This means lower profit margins. When the providers launch programs to cut costs, they typically take a hit on service. This results in dissatisfied customers, leading to cancellation or non-renewal of contracts. The need to lower prices and the need to provide a higher level of service do not go hand in hand and may soon result in third-party providers going out of business.

The second limitation is that the third-party logistics providers may not necessarily be experts in managing logistics functions. Most third-party logistics providers are part of operators of transportation networks or warehouses. They may be good operators, for instance they may be able to run trucks efficiently in a network or manage a warehouse at a location to maximize productivity, but they may not have the knowledge, experience and expertise to coordinate transportation and warehousing to minimize inventory levels and maximize customer service in terms of lead time and product variety. In other words, their core competency may not be in providing a supply chain solution through a third-party logistics service.

A third limitation is related to the marketing of the third-party logistics services. Unlike other businesses, in third-party logistics the buyer goes after the provider. Proactive marketing efforts from the providers have not been very successful. The primary reason is that there is no one department that acts as a point of contact to initiate the process from the provider side. The request for using third-party providers may stem from one or more departments within a client's organization. This is again related to the fact that logistics functions cut across multiple departments within a company. Hence, a majority of third-party logistics contracts are finalized through direct negotiations with the chief executive officers and chief financial officers.

10.3 BENCHMARKING

It is critical for a company to know the strengths and weaknesses of its business and those of its competitors in order to continue to maintain and gain market share and profits. Companies are generally aware of this during their early years but over time get more internally focused and lose sight of the environment around them. This type of insulated environment benefits the competitors since they have a better understanding of the needs of the customers. Benchmarking refers to the act of comparing a company with world-class performers and competitors involved in similar functions and operations. It is important to do benchmarking every few years since world-class performance and competitors change. A conceptual framework for benchmarking is presented in Figure 10.1 (Adam and VandeWater, 1995). The concept as illustrated in Figure 10.1 demonstrates that benchmarking identifies the gap in performance between a company and the best performer and seeks to identify ways and means to close or narrow the gap as much as possible.

Benchmarking helps an organization to identify and learn the best practices in the world for a process or a function. Benchmarking can be very useful to evaluate a company's strategies, operating plans and processes. This will help to identify flaws in strategies as well as identify processes that may require re-engineering, resulting in better practices leading to superior performance. Many big corporations and several progressive small companies use benchmarking to assess their standing against competitors in order continuously to improve their operations and processes. In the next two sections, the different types of benchmarking will be discussed and a six-step process to perform benchmarking will also be presented.

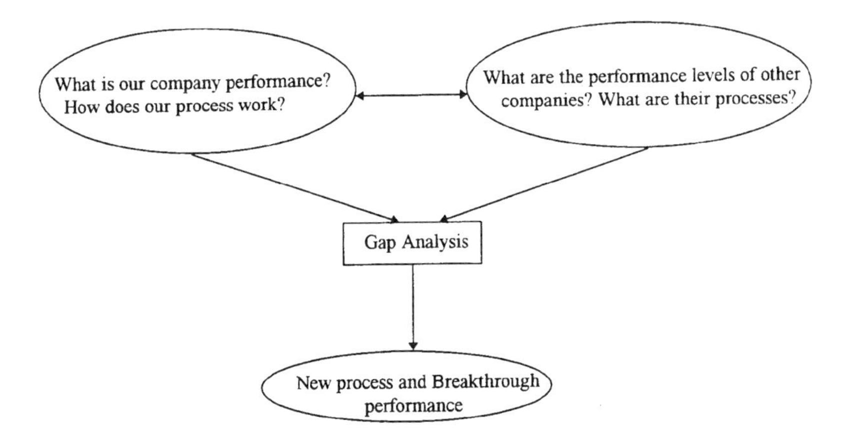

Figure 10.1 The concept of benchmarking.

10.3.1 Types of benchmarking

Benchmarking can be classified into several different types, depending upon the type of process to be benchmarked, the scope of the process and the benchmark used. For instance, a company may want to benchmark the order filling process, and the scope may be limited to the filling of in-company orders. The benchmark used may be based on world-class performance or industry-best performance. The type of process to be benchmarked may include one or more of the logistics functions or processes. Accordingly, one can have order picking benchmarking, benchmarking of truck dispatching or benchmarking of local pick-up and delivery scheduling. The scope of the process or function may include the location, type and number of in-company departments or customers. Examples include benchmarking of order picking at all the warehouses or only at the warehouses in, say, Singapore, dispatching of trucks that carry certain type of commodities to or from certain regions or the order fulfillment process for the top five customers.

Benchmarking methods based on benchmark type is shown in Figure 10.2 (Adam and VandeWater, 1995). This figure is also known as the analysis pyramid of benchmarking. This pyramid essentially lists the various types of benchmarks that can be used to compare and evaluate practices or processes. The lowest level is internal benchmarking and the highest is world-class benchmarking. Comparing the performance of a function or process with the best practice of the function or process by another internal department is internal benchmarking. A comparison of the same against the process of a world-class company is world-class benchmarking.

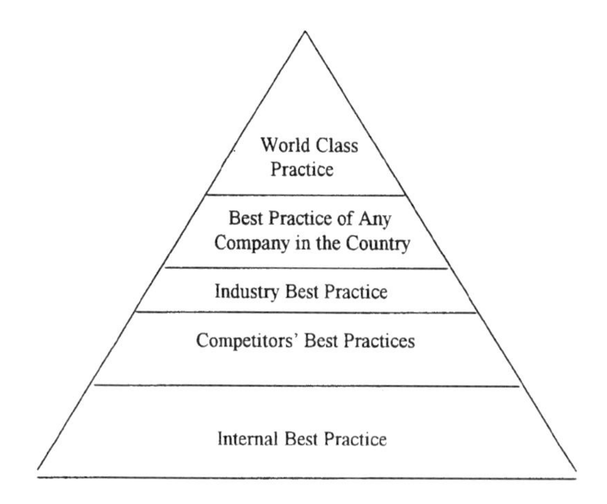

Figure 10.2 Type of benchmarking: analysis pyramid.

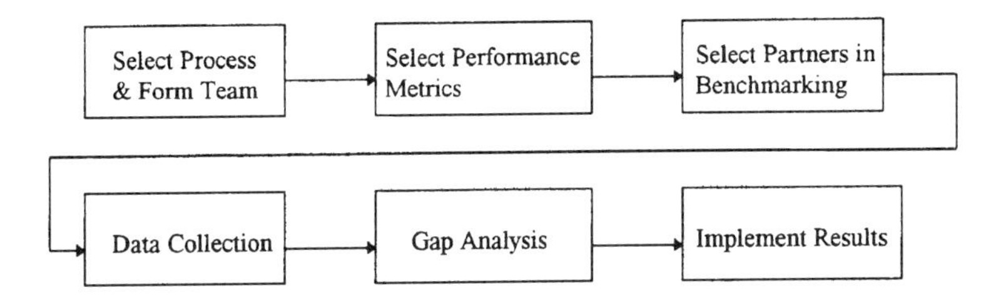

Figure 10.3 The benchmarking process.

10.3.2 Benchmarking process

The benchmarking process essentially attempts to address the following fundamental questions.

1. What should be benchmarked?
2. How should it be benchmarked?
3. What changes should be implemented based on the benchmarking results?

A six-step benchmarking process to address the above questions is graphically represented in Figure 10.3. The first step answers question 1. The next three steps are directed toward question 2. The last two steps address question 3. The six steps will now be explained in more detail.

Step 1: selection of a process and forming the team

This step really starts with the company's mission and the critical business functions to achieve the mission. Identify the processes that drive up or down the performance level of these functions. Look for functions or processes with most complaints and inefficiencies in terms of cost, quality and productivity. During this step, the company also needs to form a team that will carry out the benchmarking process. The members of the team have several functions and responsibilities – to manage the process, to act as data collectors, analysts and facilitators and to carry out other legal and clerical duties. The mix should involve people from the process and the department to be benchmarked, other similar processes and other departments involved in a similar process. It should also include people from the finance, data processing and marketing departments. The selection of the team members must be based on motivation and ability and not based on availability. The members should be trained and educated regarding their roles and be provided with the necessary resources.

Step 2: selection of performance metrics

This is very much essential since ill-defined metrics will not reveal the true performance gap between the company and the best-in-class performers. The metrics should be simple and easy to use. Several relevant metrics may be needed to analyse and compare a process from different points of view. Chapter 9 in this text book provides guidelines for selecting metrics and also provides a list of key metrics for most of the functional areas within logistics.

Step 3: select partners to use in benchmarking

This step depends largely upon step 1, which addresses the type of process, the scope of the process and the level of benchmarking. The partners in benchmarking may be a customer, vendor, competitor, another department or a subsidiary. They may be involved in processing different products and may be operating at different locations. The selection should again not be made based on availability or convenience. Several sources are available to identify partners – periodicals, trade magazines, annual reports and electronic databases. If the scope of the process to be studied and the level at which benchmarking (world-class or internal comparison) is to be done is restricted, then partners must be selected accordingly.

Step 4: data collection

Sources of data may vary depending upon the process, scope and the partners in benchmarking. Federal, state and city governments most often may maintain some of the financial statistics. Process details may be available at professional symposiums and consortiums. Certain periodicals also publish some non-proprietary statistics. The contractors of the partners may have the information as well. Surveys and telephone interviews may also be used for data collection.

Step 5: gap analysis

This step involves the comparison, by benchmarking, of performance, procedures and processes of the company with those of its partners. The first step is to understand the data available on hand in terms of its accuracy and integrity. A few checks must be made to check the validity and consistency of data. For instance, a company may report products sold in a period as 2000 units in one report and the labor hours spent as 1000 hours in another report, but the average time spent by a worker is only 20 minutes. The analysis should focus on identifying why performance is better or worse than the partners'. It should also identify how poor performance can be fixed. The concept of dominance must be used since benchmarking is typically done by using multiple metrics. An example may be to benchmark a process based on two metrics – cost and quality. When two metrics are used, a benchmarking partner is better than the company only when the partner is at least at the same level as the company in one metric and better in the other metric. The benchmarking team must produce a report outlining its findings. The findings must include flow charts of the processes, weaknesses and strengths of the processes and a comparative analysis based on the selected metrics.

Step 6: implementing the results of benchmarking

This step involves identifying the necessary actions to implement the findings of the study. One of the critical requirements for a successful benchmarking process is that the results should be actionable. If the results of benchmarking cannot be translated into actionable initiatives then the process is of very little value. First, the company must decide on the alternatives available to implement any changes. Some of the alternatives may be based on how the partners perform the process. Others may be variations of the alternatives based on the inputs from company employees through re-engineering. The alternatives must be evaluated based on the availability of resources, timing and impact. Once an alternative is selected the appropriate resources must be allocated and responsibilities be assigned in order to implement the needed improvements.

10.4 REVERSE LOGISTICS

Traditionally, logistics has been defined in two dimensions. The first dimension focuses on materials management and the second focuses on distribution. Material management covers all the functions from raw material acquisition to production at plants, and distribution covers movement from plant to customers. Reverse logistics adds a third dimension. Reverse logistics, as the name implies, deals with the handling, storage and movement of material that flows from the end customer back to the seller or supplier. This includes returns, defectives, containers or boxes and packaging material. Reverse logistics is also known as 'environmentally responsible logistics' since it helps to recycle unwanted material (boxes, bottles etc.) and recirculate returns or defectives to other stores (thrift stores, flea markets etc.). This helps to save compacting, hauling and landfill costs. The opportunities to implement reverse logistics is enormous. The average retailer and manufacturer anticipates about 5%–10% of their merchandise being returned. Catalog and shopping customers return up to 35% of their purchases. The top 100 retailers of the United States are likely to have processed some $34 billion in returns in 1997 (Eisenhuth, 1996). The statistics on boxes and bottles and packaging material is even more prominent.

10.4.1 Options for reverse logistics

What if the supplier of products is required to take back all the packaging material that is used (White, 1994)? Reverse logistics is basically how one can respond to this question. Different companies may react to this in different ways. One option is to avoid packing products individually and to transport them in bulk. The other possibility is to use returnable containers. Another possibility may be to integrate the forward and reverse flows. Packaging the products near markets in redesigned packing is another option. An ideal solution is probably to use a combination of the above. Additional options to process returns or packaging material are to return them to the vendor, send them to refurbishing centers for repairs, give them to charities, destroy them and liquidate them through auctions and antique stores.

Most companies still make logistics decisions within a two-dimensional environment focusing on costs related to materials management and distribution. It is essential to include the costs related to the third dimension – environmentally responsible logistics. This may change some of the decisions in terms of vendor selection, packaging methods and alternatives for production, handling, storage and transportation. Documented cases in the automobile industry indicate that returnable packaging can be a profitable reverse logistics strategy (Twede, 1996). However, there is also evidence that there can be financial disadvantages associated with this strategy. Studies done by German and Dutch industries show that separate infrastructure is needed for reverse logistics and that it increases the transport cost significantly. The effects of using returnable containers on environmental aspects appears to be positive – one German study of multi-trip containers showed that it reduced energy and water requirements, pollution and landfill needs. The cost of returnable containers should be based on life-cycle cost, not just based on the reverse logistics cost – it should include the complete trip of the container, the number of useful trips and the infrastructure to support returnable containers.

10.4.2 Integrated logistics

In a broad sense, reverse logistics must be incorporated into traditional logistics frameworks to formulate an integrated logistics pipeline (Dunn *et al.*, 1995). Essentially, an integrated logistics pipeline includes a flow in the reverse direction starting from the customers. The flow may be synchronized with the backhaul movement of the distribution flow for maximum efficiency wherever possible. In other words, one should schedule a pick-up of returns or packaging material from a store with the delivery of products to that store. The locations for recycling centers or the selection of recycling centers may be based on proximity to end customers where returns or packing material are generated. In other words, the design of the supply chain must include considerations related to reverse logistics such as location of landfills, incinerators and recycling centers, transportation of returns or packing material and the type of containers to store and move products.

10.5 GLOBAL LOGISTICS

The globalization of economic activity has been one of the most important and challenging developments that is having a significant impact on the industries of the developed countries. The challenge is often met in two ways by the developed countries. The first approach is to improve their processes in order to be cost competitive with third world countries which produce similar or better-quality products at lower costs. The second approach is to identify partners in other countries to manufacture components, subassemblies and, in some cases, even the final products. The selection is based on manufacturing and logistics costs for companies in various countries. The second approach forces most developed countries to get into a new area called 'global logistics'. Global logistics involves more than managing distance, currency and customs duty. It involves understanding the business and political environment of the countries involved as well as the culture and language of the workforce involved.

Some of the perceived benefits of global operations are access to lower priced raw materials and end products, better quality, increased internal competition and better customer service. Some of the disadvantages are unreliable delivery, poor communication and longer lead times. Some of the perceived challenges of global logistics are cultural and linguistic differences, duty and custom requirements, just-in-time requirements, logistics support for longer supply chains, finding qualified global suppliers or manufacturers, fluctuations in exchange rates, knowledge of foreign business practices, nationalistic attitude and behavior and understanding the political environment (Scully and Fawcett, 1994). In the following sections, an attempt briefly to address some of the challenges will be made.

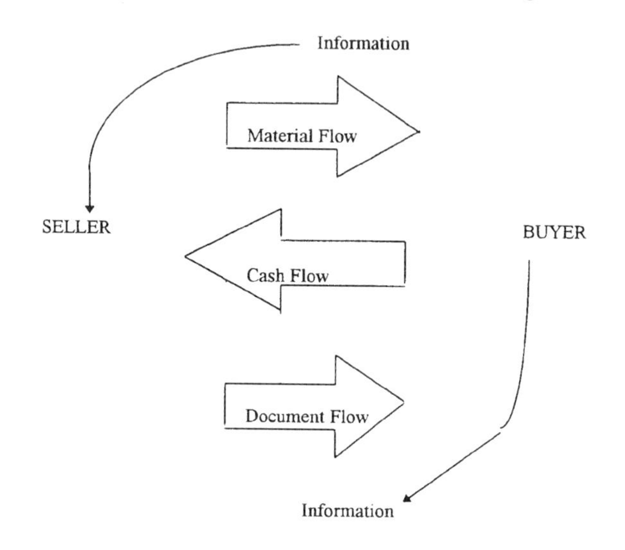

Figure 10.4 Flows in global logistics.

10.5.1 Flows in global logistics

There are three major flows involved in global logistics: material flow, document flow and cash flow. This is shown in Figure 10.4 (Rey, 1996). Material and documents flow from seller to buyer whereas cash flows from buyer to seller.

Material flow

Material flow in global logistics is very similar to that in domestic logistics except for the fact that some of the vendors, warehouses and plants are in foreign countries. This results in a larger supply chain with little control over some of its components. The biggest change may be in carrier and mode selection decisions. A global supply chain may require the involvement of other parties besides the seller and the buyer. These parties typically include trading companies, customs brokers, freight forwarders and customs and immigration authorities. Some of the common issues to be resolved between the buyer and the seller include who pays for the air freight, who is responsible for insurance and who prepares the document. For most of these issues and other questions, the answer is INCOTERMS (international commercial terms). INCOTERMS defines the mutual obligations of the seller and buyer arising from the movement of goods under an international contract from the standpoint of risks, costs and documents (Rey, 1996).

Cash flow

The basic difference in cash flow between domestic logistics and global logistics is the uncertainty associated with payment and receipt of goods arising from the involvement of different countries that may be very far apart and governed by different local regulations. The buyer does not want to pay for the goods upfront since there is uncertainty associated with receiving them. The seller does not want to ship the goods because of the uncertainty of receiving the payment. Hence a neutral party such as a bank is used to act as an interface between the buyer and seller in cash transactions. Most global transactions are carried out by using a letter of credit (LC). An LC is a financial document issued by the buyer's bank authorizing the seller to receive payment in accordance with the terms and conditions of the business transaction. After the goods are shipped to the buyer, the seller presents the LC to the bank and receives the agreed upon payment.

Document flow

There are several different types of documents involved in a global business transaction: contracts, transportation documents such as airway bills and bills of landing, invoices, cash flow documents such as letters of credit and customs clearance documents such as certificates of origin. There are standards for these documents in terms of the number of copies and the recipients, the type of information to be included and the flow of documents. Other issues include which and what copies of the documents precede the shipping of goods, which documents accompany shipping and which documents follow the shipment of goods.

10.5.2 Global logistics model

A global logistics supply chain is influenced by the governments at both ends of the transaction as well as other international institutions. Different sectors of the local governments may institute an array of guidelines, rules and restrictions. Incentives and subsidies, local content regulations and restrictions on the amount of export or local sales are very common examples. The most common restriction is the stringent inspection requirements on importing food and agricultural products from overseas. The level of customs duty on different products that enter the country is another example. Regional trade agreements such as in the European Economic Community (EEC) and the North American Free Trade Agreement (NAFTA) govern a group of countries over a certain period of time on certain commodities. There are also other international institutions such as the World Trade Organization

(WTO), International Monetary Fund (IMF) and the World Bank that facilitate and govern global logistics.

The additional rules and guidelines arising from the involvement of several agencies and the presence of additional cost variables because of the presence of neutral banks and customs agencies warrants the need for an enhanced model to address the global logistics supply chain. Conceptually, the model may be stated as follows:

$$\begin{aligned}\text{minimize } Z = \text{global logistics cost} &= \text{costs of (transportation} \\ &+ \text{warehousing} + \text{inventory} + \text{in-transit inventory)} \\ &+ \text{customs duty} + \text{taxes} + \text{insurance} \\ &+ \text{cost of other financial transactions}\end{aligned}$$

subject to capacity constraints, demand requirements, lead time and product quality and other trade or government agreements.

A global supply chain model is usually very large if one attempts to include all the details. A better approach may be to include only the high-level details – all the activities within a country may be represented as one activity, or alternatively only the entry or exit points or locations may be included in the global logistics model. This will help to formulate a model that can be solved to make the appropriate logistics decisions.

10.6 VIRTUAL WAREHOUSING

Stuart *et al.* (1995) defined the concepts of the virtual warehouse and discussed the emerging networking and communications technology that will support virtual warehousing. Virtual warehousing (VW) is a concept that will set the framework for applications that will accelerate the competitiveness of some of the large companies in the international markets. The VW concept is defined as one worldwide system that carries out dynamic and continuous material logistics functions utilizing hybrid algorithms that perform at the efficiency and accuracy levels achieved only by world-class single-location distribution centers. A conceptual model of virtual warehousing is given in Figure 10.5. The foundation of the concept is built around the emerging technologies that will allow massive amounts of data to speed around the world enhancing the accuracy of databases in a real-time environment. The advantages of such a system would be quantum gains in accuracy and throughput, on-line material visibility for customer service, precise control of transportation and data analysis capabilities to anyone capable of accessing the virtual databases. The VW concept is anticipated to reduce inventory, increase throughput and enhance customer service. The concept of virtual warehousing has still not

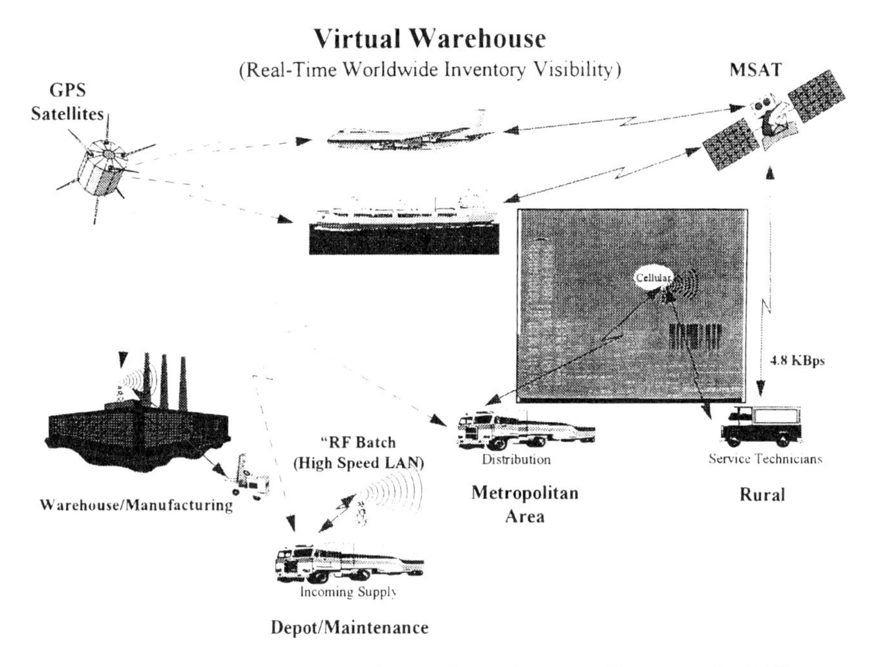

Figure 10.5 Conceptual model of virtual warehousing (Stuart *et al.*, 1995).

reached its maturity in terms of development and application. It has been implemented in pieces at some of the electronics and telecommunications companies. A fully fledged implementation requires several research and practical issues to be resolved.

10.6.1 Technology developments supporting virtual warehousing

VW is a new and emerging concept with a great potential for substantial research opportunities. The research opportunities are in terms of identifying some of the critical parameters such as stock levels at various locations, types and capacities of communication channels, the development of procedures and algorithms for material tracking and movement and the creation of innovative technologies for storage, transportation and communication. Successful implementation of VW greatly depends upon advanced communication capabilities, efficient transportation alternatives and algorithms to make stocking, communication and transportation decisions (Kasilingam *et al.*, 1996).

Portable telephones are getting smaller, smarter and less expensive. New technologies are being developed which can handle text and graphics as well as audio messages, with full video not far behind. Many of these systems are going digital, which will allow even greater accuracy and speed. Built-in radio modems will be able to link laptops, personal digital assistants and other hand-held digital devices through wireless networks. Smart cards, global positioning satellites and even triangulating signals could be used to keep track of the physical position of these different devices. Daggat (1995) also presents the benefits of satellite communication. Satellites have the advantage over land-based systems because they can provide a large area of coverage as well as flexibility. Wireless communications provide a myriad of advantages for implementing virtual warehousing.

10.7 SUMMARY

In this chapter, an attempt has been made to present an overview of some of the recent trends in logistics. There are several other developments that have taken place and are continuing to take place in a number of other areas in logistics. Some of them are very high-level concepts and others are very detailed models and algorithms related to technical issues. Some of the concepts discussed here are applicable to other areas of business as well. For instance, benchmarking can be done for any process, be it manufacturing or the design of an automobile suspension. Similarly, third-party providers may be used for the hiring of new employees or the manufacturing of wheels for an automobile company. Other concepts are specific to logistics. The concept that has a very high potential for use in the future is probably virtual warehousing. The emergence of innovative technologies in the areas of communication, tracking and scanning coupled with the increasing trend in global partnerships will make virtual warehousing cost-effective and an absolute necessity in the next century. The scope for further research and development is plenty for reverse logistics, global logistics and virtual warehousing.

PROBLEMS

10.1 Briefly explain the third-party logistics provider selection process with use of an example where the company is in the business of providing oil and gasoline products to its 200 retail pump outlets.

10.2 Why do companies benchmark? Give an example each for the various types of benchmarking.

10.3 What are the options available to process returns or package material? Discuss the relative merits and demerits of the options

10.4 Explain briefly the barriers to implementing virtual warehousing.

10.5 Discuss the additional dimension or the flow in global logistics. What is its impact on the design of the supply chain?

REFERENCES

Adam, P. and VandeWater, R. (1995) Benchmarking and the bottom line: translating business reengineering into bottom line results. *Industrial Engineering*, 24–6.

Armstrong, R.D. (1996) Who's who in logistics? Primary features of the major players. *Annual Conference Proceedings*, 625–34.

Canitz, H.C. (1996) How to find, evaluate, choose, and manage a third party logistics provider. *Annual Conference Proceedings*, 597–624.

Daggatt, R. (1995) Satellites for a developing world: satellites could provide universal access to the information economy. *Scientific American*, 94.

Dunn, S.C., Wu, H.J. and Yound, R.R. (1995) Environmentally responsible logistics: an academic view. *Proceedings of the 24th Annual Transportation and Logistics Educators Conference*, 77–104.

Eisenhuth, D. (1996) Asset recovery: adding value to assets. *Annual Conference Proceedings*, 573–7.

Kasilingam, R.G., Landers, T.L. and Walker, B.H. (1996) A framework for designing virtual warehouse. The Logistics Institute Report, The University of Arkansas, Fayetteville, AR.

Maltz, A.B. and Lieb, R.C. (1995) The third party logistics industry: evolution, drivers and prospects. *Proceedings of the 24th Annual Transportation and Logistics Educators Conference*, 45–76.

Rector, P. (1996) Alternative asset recovery. *Annual Conference Proceedings*, 569–72.

Rey, M.F. (1996) Global logistics: surviving the minefield. Technical report, The Logistics Institute at Georgia Institute of Technology, GA.

Scully, J.I. and Fawcett, S.E. (1994) International procurement strategies: challenges and opportunities for the small firm. *Production and Inventory Management Journal*, 39–46.

Stuart, D.E., Owen, J. and Landers, T.L. (1995) Establishing the virtual warehouse. The Logistics Institute Report, The University of Arkansas, Fayetteville, AR.

Twede, D. (1996) Do returnable containers for large finished goods make sense? Returnable packaging considerations. *Annual Conference Proceedings*, 583–7.

White, J.A. (1994) Reverse logistics moves forward. *Modern Material Handling*, 29.

FURTHER READING

Blumberg, D.F. (1994) Strategic benchmarking of service and logistics support operations. *Journal of Business Logistics*, 89–119.

Hitchcock, N.A. (1993) Benchmarking bolsters quality at Texas Instruments, Modern Materials Handling, 46-48.

Lieb, R.C. and Randall, H. (1996) CEO perspectives on the current status and future prospects of the third party logistics industry in the United States. *Annual Conference Proceedings*, Council of Logistics Management, 401–30.

Lieb, R.C. and Randall, H. (1996) A comparison of the use of third party logistics services by large American manufacturers, 1991, 1994, 1995, and 1996. *Annual Conference Proceedings*, 431–51.

Pulat, M. (1994) Benchmarking is more than organized tourism. *Industrial Engineering*, 22–3.

Scully, J.I. and Fawcett, S.E. (1994) International procurement strategies: challenges and opportunities for the small firm. *Production and Inventory Management Journal*, 39–46.

Venetucci, R. (1992) Benchmarking: a reality check for strategy and performance objectives. *Production and Inventory Management Journal*, 32–6.

APPENDIX A

Logistics and transportation cases

Case study 1, Saturn Corporation: improving the plant–retailer link in the auto industry supply chain – a logistics case study developed by Gibson, B.J. and Wilson, J.W. for the Council of Logistics Management, Oak Ridge, IL.

Case study 2, Toys 'R' Us – a logistics case study developed by Kay, M.J. for the Council of Logistics Management, Oak Ridge, IL. Both case studies are reprinted with permission from the Council of Logistics Management.

A.1 CASE 1. SATURN CORPORATION: IMPROVING THE PLANT–RETAILER LINK IN THE AUTO INDUSTRY SUPPLY CHAIN

This case focuses on the transportation of finished vehicles from Saturn's Spring Hill, TN, assembly plant to Saturn retailers in the United States. The case provides insight into the delivery process, the objectives of the process stakeholders and performance of the current system. All three issues play key roles in the task at hand – the evaluation and selection of appropriate delivery methods to enhance the performance of the Saturn supply chain.

The primary purpose of the case is to help the reader understand the complexity and implications of carrier selection and evaluation. The case introduces the reader to an industry where delivery cost minimization has been the predominant strategy. Then, the reader is presented with a scenario in which a broader supply chain strategy is used to address the needs of the organization (Saturn), channel members (Saturn retailers) and consumers (car buyers). Throughout the case, the reader is presented with relevant information regarding cost, customer service, shipment integrity and control. This information must be considered jointly to conduct an effective case analysis. A key goal of the case is to encourage the reader to look beyond the obvious transactional costs of delivery and make decisions that consider all relevant costs, balance costs and service levels and address the requirements of the process stakeholders.

The scope of the case involves the entire delivery process (from the time a vehicle comes off the production line until it is physically put in the retailer's showroom). However, given the case objectives and the size of the retailer network it is impractical to consider the entire system. Therefore the scope of the case is limited to a specific group of new Saturn retailers. The geographic dispersion of this group provides a variety of viable strategies. In some locations, truck delivery has an advantage over rail delivery and vice versa. In other locations there is no clear-cut answer. In those situations the reader must delve into qualitative issues and resolve the conflicting stakeholder requirements. That alone will help the reader gain a greater appreciation for the complexity of this strategic decision.

Overall, the case should foster a better understanding of logistics in the automobile industry, carrier selection and evaluation processes, as well as supply chain management concepts. In group settings the case issues should promote a lively discussion regarding the key issues, stakeholder preferences and appropriate courses of action.

A.1.1 Company overview

In 1982 General Motors (GM) began formal work on the concept for a new small car that could compete head-to-head with the most successful foreign imports. For years, the US auto industry had been losing ground to the high-tech, high-quality, reasonable price image of several foreign auto manufacturers, primarily the large Japanese firms. The new-product concept team at GM was given the task of designing a small, distinctive looking high-quality automobile that would exemplify leading-edge technology in design, manufacturing and distribution.

GM did not want to develop a vehicle that could temporarily regain a few market share points. GM desired a new vehicle and a new development process that would yield an innovative company, one that could produce vehicles quickly, operate efficiently, provide dependable quality and be responsive to the needs of both the parent company and the market place. It was further hoped that this new car and company concept would lead to the reorganization of all GM divisions, allowing them to regain their long-term competitive position both in domestic and in foreign markets. Quite a formidable mission!

Start-up and mission

The end result of the planning process, the Saturn Corporation, came into being in January 1985. Saturn was formed as a wholly-owned subsidiary of GM. In order to insure the ability of Saturn to be innovative and creative the company was given complete operational independence from the parent company. This was a significant departure from GM's rigid corporate structure and represented a considerable risk for the entire GM corporate family. The initial financial risk was formidable as well. GM allocated $5 billion to the Saturn project, proving its commitment to continued leadership in the US automobile industry.

After an extensive and very highly publicized search for a suitable location for the plant was conducted, Spring Hill, TN, was chosen. Spring Hill, which at the time of selection had a total population of only 1100 people, is just 30 miles south of Nashville and within minutes of Interstate 65, Interstate 40 and Interstate 24. The site is also within 600 miles of approximately 60% of the nation's population. Although the state of Tennessee did not grant Saturn any significant tax concessions, the state did agree to provide $50 million for employee training and highway improvements. About $30 million of this total was used to construct Saturn Parkway, a four-lane divided highway connecting the plant and facilities with Interstate 65. This project was completed in 1989. Rail service to and from the site is provided by CSX via a rail spur constructed in the Spring of 1988. The proximity to large numbers of potential customers and the relative abundance of transportation options appear to have influenced the decision to locate at Spring Hill significantly.

Saturn's initial mission was to 'market vehicles developed and manufactured in the United States that are world leaders in quality, cost, and customer satisfaction through the integration of people, technology, and business systems.' The mission was given substance when construction began on the production facilities in 1988.

Production operations

The 4.3 million ft^2 Spring Hill production complex was officially completed in June 1990 at a cost of approximately $1.9 billion. The first Saturn automobile was produced in July 1990. The Saturn complex is unlike any other automobile production facility in the United States. The manufacturing process is managed as an integrated system from the source of raw materials to the final customer. The intent is truly to have customer needs and desires drive the entire Saturn operation. The highly automated manufacturing and assembly complex consists of three major operations: power train, body systems and vehicle systems. In the power train plant engines and transmissions are actually produced on-site. The second operation, body systems, includes the stamping plant, body fabrication and the paint shop. In the final vehicle systems operation, cars are assembled, tested and prepared for shipping. The vehicle systems facility is strategically located between the power train plant and the body systems plant to improve material flow.

Current production capacity at the plant is approximately 1100 units per day and the plant operates two 10 h shifts, 6 days per week. As of May 1995 there were 9234 people employed at the facility.

Logistics and marketing

The just-in-time method of inventory control is employed in order to minimize in-plant and in-transit inventory. As a result, Saturn relies primarily on truck transportation and an excellent highway system to reduce order cycle times and inventory costs and to keep materials and components arriving only when needed. When a new Saturn vehicle exits vehicle systems it is inspected and staged for loading either in the truck or rail loading area. Since each vehicle is produced to fill a specific customer or retailer order, the final destination is known before the vehicle reaches the transportation staging area.

Another highly innovative aspect of the Saturn corporate philosophy concerns the final link with the customer – the automobile dealer. The Saturn retail strategy is called the market area approach. Saturn 'dealers' are referred to as 'retailers'. Each retailer has a well-defined market area. Inventory is typically much lower than for other popular brands. This, in part, is a result of consistently strong demand and Saturn's integrated information exchange system that gives the retailer instant access to production scheduling information that can be very helpful in completing a customer order and/or handling a customer inquiry concerning delivery dates. Although retailers are not given specific delivery dates, they are given the scheduled date of production for each vehicle that is ordered.

Retail prices are set by Saturn, not by the retailers. In effect, it is a 'no-haggle' one-price policy. Every customer is treated the same and pays the same price for identically equipped vehicles. Saturn customers appear to prefer this low-pressure selling situation. They also appreciate the special attention that is given to the first-time Saturn buyer. When a first-time buyer picks up their new vehicle a picture is taken and the entire sales staff is on hand to wish the new owner well. – Not the type of treatment that US car buyers are accustomed to receiving at traditional dealerships!

In 1991, after only one year of production, Saturn ranked number one in new vehicle sales per retail outlet, averaging 776 units for the year. This was the first time that a domestic brand had achieved this honor in 15 years. In 1992 Saturn averaged 1072 new vehicle sales per retail outlet, well ahead of number two Honda with only 654. As sales per outlet have grown, so has the total number of outlets. Currently, there are 326 domestic retail outlets and 90 foreign retail outlets (70 in Canada and 20 in Taiwan).

Saturn's success is dependent upon meeting and exceeding customer expectations on an ongoing basis, given the intense competition between brands in the small-car market. There is evidence that the company is up to the task. Regular surveys of Saturn owners consistently indicate that 97% of them would 'enthusiastically recommend the purchase of a Saturn' to other potential buyers. Saturn also excels in the annual J.D. Power and Associate Customer Satisfaction Index, the automobile industry's benchmark survey. Since 1992 Saturn has been the highest ranked domestic nameplate. In 1995 Saturn achieved a record score in the J.D. Power Index and was rated as the best overall line (domestic or foreign) in sales satisfaction.

Production and sales have climbed steadily. Demand is keeping production at or very near plant capacity. Demand has been so strong, in fact, that Saturn achieved record sales in 11 of 12 months during 1994. Total calendar year sales were 286 003 vehicles. This ranks Saturn among the market share leaders in the small-car segment of the automobile industry. A milestone was reached on 1 June 1995. On that date, employees and dignitaries celebrated the completion of the one-millionth Saturn automobile. The future looks bright with the introduction of significantly restyled models for the 1996 and 1997 model years.

A.1.2 Business situation

Saturn's logistics system has always been viewed as a critical component in the overall success of the organization. The system's supply chain management capabilities are fundamental to the company's ability to operate a just-in-time manufacturing system. The logistics system focuses on three primary activities – inbound material flow, finished vehicle transportation and service parts support. The first two components fall under the responsibility of the Director of Materials Flow and Logistics. The Saturn Service Parts Organization is managed as a separate function.

Innovative logistics practices

From its conceptualization, the inbound logistics process has been touted as innovative and bold. Saturn was among the leaders in developing a strategic alliance with a single, dedicated logistics service provider. This relationship with Ryder Dedicated Logistics provides Saturn with a closed-loop, continuous replenishment system. This world-class just-in-time system allows Saturn to maintain minimal inventories (2–8 hours of production volume) since materials are received in frequent, small volumes.

The outbound logistics process uses a combination of truck and rail transportation to deliver the finished product to Saturn retailers. Retailers within a 500 mile radius of the plant typically receive vehicles via truck delivery, and the remaining retailers receive vehicles via rail. The proportion of rail to truck deliveries is roughly 60/40. Autohauler trucks and trilevel railcars are used to move Saturn vehicles. Although the outbound logistics process has not received the same attention and accolades as the inbound process, Saturn also employs many innovative practices for finished vehicle transportation. The Saturn Vehicle Transportation Team takes a much more active role in the delivery process than most automobile manufacturers do at their assembly plants.

Many automakers take a hands-off approach toward the vehicle delivery process. After a vehicle is produced and released by the plant the vehicle becomes the sole responsibility of the carrier. In contrast, Saturn's structure of responsibility is far less rigid. All parties involved in the outbound process are expected to share information and resolve problems collectively before vehicles are dispatched from Spring Hill. The ultimate goal is to ship perfect vehicles to Saturn retailers. Another difference between Saturn and most automakers is the staging and loading facility location. As automakers have turned more outbound responsibilities over to third parties, the number of plant-side loading and dispatching facilities has decreased. Saturn's on-site staging and loading facilities are located directly behind the assembly plant where the process can be easily monitored. The vehicle shipping office is located amongst these facilities. This direct contact with the operation fosters increased process knowledge, frequent interaction with the outbound service providers, real-time problem-solving capabilities and improved communication.

The third unique aspect of Saturn's outbound logistics process is their relationship with Quality Automotive Transport (QAT), a subsidiary of Ryder Corporation. QAT serves as a dedicated contract carrier for Saturn, moving more than 80% of all truckload shipments of vehicles. QAT also manages the marshaling yard and provides traffic management services, including coordination of Premier Services and CSX activities. Premier Services provides plant to staging area vehicle transfers and rail loading services. CSX is the origin carrier for all outbound rail shipments. This relationship provides Saturn with a single point of contact regarding outbound activities. Few other automakers have established an alliance of this magnitude with a single outbound logistics provider.

These innovative practices have produced excellent results for Saturn, its retailers and its customers. Saturn is among the GM leaders in outbound transportation performance, including highest percentage of vehicles dispatched within 72 h of production (the corporate standard) and lowest frequency of post-production damage. In fact, Saturn's

Table A.1 Order processing event codes

Event Code	*Event Description*	*Responsibility*
10	Retailer order is developed	Marketing
25	Retailer order is submitted for production	Marketing
33	Vehicles are sequenced for production	Production control
38	Vehicle is produced	Assembly Plant
40	Vehicle is available for shipment	Assembly Plant
42	Vehicle is loaded and dispatched	Carrier
48	Vehicle is unloaded and made available for drayage	Carrier
50	Vehicle is delivered to dealer	Carrier

vehicle transportation team has received the Vehicle Transportation Quality award in 1994 and 1995. This award is presented by GM to only the highest performing facilities (internal and external) that load, unload and/or transfer vehicles.

Order process

The production and delivery of Saturn vehicles emulates the GM order processing system and uses GM event codes to control the process. Table A.1 contains all event codes relevant to the vehicle delivery process. In this system, the Marketing Department develops a retailer forecast based on past sales (event 10) approximately six weeks prior to production. This preliminary order is communicated to the retailer who has up to three weeks to make major changes to the order (e.g. place a special customer order). Minor order modifications (e.g. add a sun roof, change the stereo etc.) can be made up to one week prior to production. The final order is then submitted to Production Control (event 25).

The retailer order is accepted by Production Control, then put into production sequence (event 33) three days prior to assembly. When the vehicle is sequenced, no further order changes are allowed. At this time, unique vehicle identification numbers (VIN) are assigned to each vehicle in the retailer order. The VIN is used throughout the Saturn system to track a vehicle from production to customer sale and beyond. The window stickers are also created at event 33 to provide information on product features, price and retail delivery location. All Saturn vehicles are produced to specific order, not to general or pool stock.

Delivery process

The Saturn delivery process also begins at event 33. At sequencing, a copy of the order record is sent via electronic data interchange to GM's logistics information system. The system identifies the preferred mode (truck or rail) and route for retail delivery. This information is transferred electronically to QAT in the form of an advance shipping notice (ASN). Using this information, QAT builds a file for each vehicle and begins to develop a general delivery plan.

Truck delivery planning focuses on the anticipated availability of QAT drivers and equipment. QAT personnel analyse the ASNs by retailer to determine the number of trucks and drivers needed. Each QAT truck can carry 12 Saturn vehicles. Under ideal conditions, 12 vehicles are sequenced for a particular retailer, produced within a 3–4 hour window and released to QAT for a one-stop delivery. If it appears that multiple stop loads are imminent, the QAT supervisor plans economical delivery combinations. Other Ryder carriers are used when QAT has inadequate capacity or when a cost-reducing backhaul opportunity is present.

Rail delivery planning focuses on the number of rail cars needed in Spring Hill. Using the sequencing information, QAT personnel analyse the number of vehicles destined for a particular rail ramp in the GM network. The total is divided by 15 (the number of Saturns that fit on a trilevel railcar) to determine the number needed. This analysis is conducted for each of the rail ramps in the system. The overall forecast is sent each Thursday morning to GM's central logistics operation in Detroit, MI, which then orders rail cars from CSX.

After each vehicle is produced (event 38) it is put through emissions testing and a final Saturn employee inspection at the 'Inspiration Point'. If everything is acceptable the vehicle is released to the carrier (event 40). A Premier Services employee scans a bar-coded card of the VIN into the QAT computer system and drives the vehicle out of the assembly plant. The card also contains vehicle destination and delivery method information.

Truck delivery If the vehicle is scheduled for truck delivery it is driven to the truck marshaling or loading yard. The vehicle is parked in a lane that has been designated for the destination retailer. The lanes are clearly marked with numbers as are the 12 spaces within the lane. Premier employees are trained to park the cars in a consistent fashion which provides a cushion of space around the vehicle. Uniform staging helps to reduce the incidence of bumps, scratches and paint chips from careless handling. After the vehicle is staged, the VIN is checked by a QAT supervisor for accuracy. This inspection helps prevent vehicles from being misloaded and delivered to the wrong retailer.

The driver is responsible for inspecting all 12 vehicles for damage and verifying the accuracy of the VIN on each vehicle. The QAT supervisor must investigate and resolve any problems that the driver discovers in the inspection process. If necessary, vehicles are restaged in the correct lane, minor repairs are made on the spot or the vehicle is returned to the plant. Next, the driver loads and secures the vehicles on the trailer. This process takes approximately 90 minutes. After the last vehicle is placed on the trailer the QAT supervisor makes a final inspection of the load. If everything checks out, the driver is given approval to trade the loading documents for freight bills in the shipping office.

After obtaining the freight bills, the driver and load are dispatched (event 42). The driver makes final trip preparations and departs for the retailer(s) on the route. Barring problems, the driver arrives at the retailer within 24–48 hours (depending on the distance and legal driving hours available). Vehicles can be delivered to most retailers Monday through Friday from 7 a.m. until 5 p.m., though some are more flexible. The driver is responsible for presenting the freight bills to the retailer representative, unloading each vehicle and conducting a joint inspection of the vehicles. All damages and problems are noted on the delivery receipt and freight bills. If necessary, a freight claim is filed in the GM logistics information system. The retail delivery process concludes when the retailer representative signs the freight bills, indicating receipt of the vehicle (event 50).

Under most conditions, the driver returns directly to Spring Hill empty for another load. Until late 1995 the QAT/Transportation Unlimited contract with the Teamsters did not contain provisions for handling automobile backhauls. This has been changed and QAT is working to reduce the number of empty miles. Overall, QAT provides delivery service to Saturn retailers at an average cost per mile of $0.39. Less than 0.5% of all QAT delivered vehicles incur damage. In addition, more than 98% of all loads are delivered on time, with an average transit time under 1.5 days.

Rail delivery If the vehicle coming off the line is scheduled for rail delivery it is taken to the rail marshaling or loading yard. The vehicle is parked in a designated lane based on its destination rail ramp (rather than by retailer). These lanes are assigned by a Premier supervisor on a daily basis. Parking protocols are similar to those used in the truck staging area. CSX is responsible for delivering an adequate supply of trilevel railcars to the Saturn plant from their Radnor yard in the Nashville area. CSX contracts with two companies to assist them in the preparation and dispatch of Saturn loads. One of the contractors (Inter-rail) inspects the empty railcars for damage and other problems (e.g. missing tie-down straps) then makes necessary repairs prior to placing the railcars on the four spur lines at Saturn. They also prepare the railcars for loading by opening the doors and securing the ramps between each railcar.

The second contractor, I.T.S., has five inspectors at Saturn to monitor vehicle quality in the staging area prior to the loading process. Any damages or problems found are referred to the QAT yard supervisor for resolution. Minor repairs are made on the spot. In the rare case of a major problem the vehicle is returned to the plant. The actual loading process is performed by a special crew of Premier employees who adjust the loading ramps and drive the Saturn vehicles onto the railcars, starting with the top level. Normally, 4–5 railcars are loaded at a time on a given spur line. The loading ramps are then lowered to the second level, the railcars filled with vehicles and so on with the first level.

Since all rail cars on a single spur line may not be destined for the same rail ramp, correct loading sequence is critical. It is extremely difficult and time consuming to unload vehicles that have been improperly placed on a railcar. After the block of railcars is completely loaded a Premier employee secures all vehicles to the railcar by using heavy nylon wheel straps. These straps allow the vehicles to ride naturally using their own suspension to absorb the sway and vibration of the rail system. This is a much more effective means of securing the vehicle than the traditional method of chaining its frame to the railcar, which led to high levels of suspension and frame damage.

A second inspection of the vehicles is conducted by I.T.S. after the loading and tie-down process is complete. They check for vehicle damage created in the loading process, proper spacing between vehicles and proper chocking. Premier is required to correct any spacing and chocking problems. When all inspections are complete I.T.S. signals QAT that the load is ready. QAT them confirms the railcar number and destination combination against the vehicle VINs on the railcar. If everything checks out, QAT develops freight documents and advises Inter-rail to prepare the railcars for dispatch (remove ramps between railcars and close all railcar doors). QAT also informs CSX that the block is ready to be dispatched (event 42) and moved to the Radnor yard for line haul movement.

At the Radnor yard, CSX classifies the Saturn-filled railcars with other railcars by destination rail ramp and dispatches them as scheduled. CSX handles the line haul portion of the move for vehicles with southeastern destinations. In other cases, the destination is outside the CSX system and railcars must be transferred to other railroads at major interchange points (e.g. Memphis for Texas-bound loads). Upon arrival at the destination rail ramp, the vehicles are unloaded and inspected. If necessary, damage reports are filed.

The vehicles are then released to drayage carriers who inspect the vehicles, reload them onto trucks and dispatch them for final delivery to Saturn retailers (a second event 42). Drayage drivers are required to deliver vehicles according to the same schedule and processes used by QAT. Again, the retail delivery process concludes when the retailer representative signs the freight bills, indicating receipt of the vehicle (event 50). Overall, CSX provides a competitively priced rail service for long-distance deliveries to Saturn retailers. Saturn's average cost per mile is $0.32 for rail service. Less than 1% of all vehicles handled by CSX or its drayage partners incur damage. However, on-time performance and transit-time average suffer as distance increases. Less than 80% of all rail loads are delivered on time and transit time averages more than 7.5 days.

Delivery control and coordination mechanisms Managing the flow of vehicles from plant to retailers requires a coordinated effort from a variety of organizations. Saturn relies heavily on QAT to manage the day-to-day functions of inventory control, load planning and dispatching, documentation and report generation. The GM logistics information system provides key inputs to the QAT planning process and maintains a database of order processing event histories for every Saturn vehicle in the transportation system. In-transit vehicle tracing is provided by CSX, destination railroads and drayage companies involved in the rail delivery process. Retailers have a limited role in the control process.

The strongest control mechanisms exist for on-site activities. The combined capabilities of the QAT system and GM system allow Saturn easily to monitor the status of vehicles before they are dispatched. QAT generates daily reports that indicate the number of vehicles shipped (by mode), the number of loads shipped (by carrier – CSX, QAT or backhauler), the number of vehicles waiting to be shipped (by mode) and the number of vehicles shipped within 72 hours. QAT also knows the age of each vehicle that is in the marshaling yard.

In-transit control is more challenging than on-site control. Saturn depends on its carriers to monitor shipment progress and safety, provide tracing services and to communicate effectively. These challenges increase with distance and the number of carriers involved. Truck shipments are relatively easy to monitor in-transit because a single carrier is moving Saturn vehicles less than 500 miles. Rail shipments are more difficult to control. The increased distance and involvement of 2–4 carriers expands the opportunity for delays, damage and other problems. Also, the rail carriers' tracing capabilities are not yet on a par with those of motor carriers.

Retailer involvement in the process is typical of a free-on-board (FOB) destination delivery. They schedule and control the unloading process, inspect the vehicles and take ownership. However, they have no control over the mode or carriers used, shipment or delivery dates or number of vehicles shipped at a time. They receive no ASNs or delivery date forecasts but are aware of the event 40 and 42 occurrences. This allows them to make rough estimates of when loads can be expected.

Table A.2 New dealer service forecast

Retailer Location	*Distance from plant*	*Transit Time*		*Linehaul Cost Per Load*	
		Truck	*Rail*	*Truck*	*Rail*
Baton Rouge, LA	466	1	6	2405	2796
Chicago, IL	473	1	8	2440	2838
Savannah, GA	508	1	7	2560	2935
Kansas City, MO	538	1	5	2680	3079
Jacksonville, FL	606	2	6	2890	3272
Dallas, TX	668	2	7	3166	3530
Omaha, NE	722	2	7	3335	3680
Watertown, NY	1012	3	9	4100	4300

A.1.3 Problem definition

Saturn's popularity and record sales are driving the need for more retailers. This year the Marketing Team has developed 20 new market areas where Saturn retail facilities will be located. These facilities will be opened throughout the year. There are a variety of logistical issues to address before a new retailer holds a grand opening. However, the most pressing and challenging decision is whether to transport vehicles by truck or rail. Once this decision is made, all other outbound transportation issues can be readily handled.

The Vehicle Transportation Team has meticulously analysed the delivery options for each retail location. Their efforts have revealed that truck delivery and rail delivery have compelling advantages for five and seven of the new sites, respectively. That leaves eight sites without an obvious decision. Both delivery methods offer some advantages for these sites, as highlighted by Table A.2. The team must weigh the advantages and disadvantages of the two methods against stakeholder concerns and make a vehicle delivery decision for each site.

Saturn Corporation concerns

Saturn has two overriding interests – to be a profitable manufacturer of small automobiles and to be recognized as a 'different kind of car company'. To achieve profitability, every activity must be performed in a cost-effective manner. To be truly different, every portion of the organization must stress customer satisfaction. To achieve both is extremely difficult. These interests translate into a major challenge for the Vehicle Transportation Team. The Team must strive for cost control and continuous improvement of performance. Key performance issues include cost control, damage rates, single dealer loads and inventory turnover.

Cost control
Cost control is of the utmost importance. In the automobile industry, inbound and outbound logistics costs are passed along to the customer in the form of destination charges. Owing to the low price of Saturn vehicles this destination charge has a much greater impact on the overall purchase price than would the destination charge on a Cadillac. Thus the Vehicle Transportation Team is inclined to use low-cost methods of retail delivery to lessen the logistics impact on the retail price.

Damage rates
Minimization of transportation-related vehicle damage is an ongoing issue. Although auto manufacturers and their carriers have vastly reduced damage incidents and costs from the levels experienced 10 years ago, opportunities for further reduction exist. Damage prevention rather than inspection needs to be the priority, according to the Vehicle Transportation Team.

Inspections only serve to pinpoint who caused damage to a vehicle. This leads to accurate determination of who should pay for a damage claim but does not necessarily indicate why the damage occurred or prevent future incidents. On the other hand, increased training of vehicle handlers (at the plant, carrier and retailer levels) and decreased vehicle handling (unloading and reloading at transfer points) reduce the likelihood of damage. Thus the Vehicle Transportation Team prefers to work with carriers that stress employee training and whose processes minimize vehicle transfers.

Single-dealer loads
Increasing the proportion of single-dealer deliveries is another priority. Saturn has found that 12 unit, single-dealer truckloads result in fewer in-transit problems and less handling-related damage than do multistop loads. Other positive results include increased equipment utilization and improved shipment tracking capabilities. These results translate into lower operational costs for carriers (i.e. QAT, backhaulers and drayage companies), fewer delivery delays and receiving headaches for retailers and fewer distractions for the Vehicle Transportation Team.

Inventory turnover
Accelerating the flow of inventory from the marshaling yard is the other internal priority. Saturn boasts nearly 100% compliance with the 72 hour guideline for dispatching new vehicles. However, a Saturn benchmarking study revealed that a few automakers dispatch their vehicles within 24 hours of production. The Vehicle Transportation Team has a strong desire to achieve comparable results, though faster inventory turns may require more multiple dealer loads.

Saturn retailer concerns

In order to succeed at the retail level the right assortment of available vehicles is needed. It is much easier for the Saturn sales consultants to demonstrate and sell vehicles when the customer's preferred model–color–options combination is on hand. After all, the customer cannot test drive a photograph. Thus the retailers desire fast, frequent replenishment of popular models. Although much of the responsibility for the replenishment of vehicles is outside the control of the Vehicle Transportation Team they have a role in the process. By the time event 40 occurs, retailers are anxious to receive the vehicles for two reasons – costs minimization and customer service.

Transit time and dealer costs

From a customer service standpoint the faster the vehicle arrives at the retail outlet the easier it is to sell. The retailer cost impact is less apparent, but no less important. In the Saturn system retailers are invoiced for vehicles at event 42 (vehicle is dispatched) and automatically pay Saturn via electronic funds transfers 14 days later. Dealers in other systems do not have to pay for 60 days. Obviously, this is a critical issue to Saturn retailers. They desire fast transit times to maximize their opportunity to sell vehicles during this short 'grace period'. Prevention of vehicle damage is also important as repair activities delay the opportunity to sell vehicles (not to mention the negative impact on customers' quality perceptions).

Customer service and special orders

The other major retailer issue concerns the special customer order. Although customers will accept a projected delivery date of six weeks at the outset, their tolerance for estimates decreases as time elapses. Retailers would like to be able to provide the special order customer with firm, dependable delivery dates. Advanced shipping notification, transit time consistency and rapid shipment tracking capabilities (e.g. satellite tracking) would enhance the retailers' ability confidently to commit to firm delivery dates.

The current challenge

In terms of vehicle delivery process preferences, the retailers' goals do not always mesh perfectly with the goals of the Saturn organization. Developing solutions which effectively consider all issues and appease the stakeholders is the challenge. The Vehicle Transportation Team must select delivery methods for each new retailer that provide the best possible balance of cost and service.

A.1.4 Analysis and alternatives considered

In order to make effective selection decisions the performance of the two delivery methods is monitored continuously by the Vehicle Transportation Team in coordination with QAT. Most of the information is collected manually, with the balance of data coming from the GM logistics information system. Performance analysis focuses on three key areas: cost, damage and transit time.

Cost of service

The cost of delivery is analysed on a per vehicle basis and includes all of the expenses incurred by Saturn. The total cost includes yard activities, line haul moves, drayage and other related expenses. Rail and truck delivery costs are compiled independently, then compared. Truck delivery costs are relatively easy to capture since QAT provides the vast majority of deliveries. Transfers from the assembly plant, inspections and marshaling activities conducted by Premier employees average $2.50 per vehicle. All other loading, line haul, dealer delivery and traffic management activity costs are included in the truckload rates negotiated by Saturn with QAT. These rates are based on the distance from Spring Hill to the retailer. On average, this cost is $0.39 per mile for each vehicle with a range of $0.29–$0.45 per mile.

Rail delivery costs are more difficult to assess because of the number of organizations involved in the process. The yard activities conducted by Premier employees (vehicle transfers from the assembly plant to the marshaling yard and railcar loading) cost Saturn approximately $8.50 per vehicle. Each I.T.S. inspection of a vehicle costs $0.90. Rail line haul cost per vehicle averages $0.32 per mile, with a range of $0.22–$0.42 per mile. Drayage from the destination rail ramp to the retailers runs at an average of $12 per vehicle. Table A.3 provides a comparison of estimated truck costs with rail costs for four current retailers. It reveals that rail holds a distinct advantage over truck as distance increases. In all but a few unusual situations, Saturn's transportation rates increase at a decreasing rate.

Table A.3 Typical retail delivery costs

Plant–Retailer Distance (miles)	*Truck Delivery Costs ($)*		*Rail Delivery Costs ($)*	
	per mile	*per vehicle*	*per mile*	*per vehicle*
250	0.45	115.00	0.42	125.50
500	0.42	212.50	0.38	210.50
750	0.38	287.50	0.33	260.50
1000	0.34	342.50	0.29	310.50

Vehicle damage

Vehicle damages related to the delivery process are evaluated on four different bases to provide a comprehensive picture of the problem. The proportion of transportation claims per vehicles shipped indicate the frequency of vehicle damages. The cost per transportation claim measurement highlights the financial magnitude of vehicle damages. Both measurements are analysed on a modal basis. The breakdown of claims by damage type and by responsible party often provide valuable insight for process redesign and damage prevention efforts.

Overall, Saturn's performance in this area is exemplary. Less than 1% of all vehicles shipped to retailers are subject to retailer transportation claims. Only 0.45% of all truck deliveries incur damages. In contrast, 0.92% of all rail deliveries incur damage. However, the financial impact of claims is reversed. An average damage claim for truck deliveries is $204, whereas the average damage claim for rail is only $119.

The majority of claims are cosmetic in nature and do not affect the performance of the vehicle. Paint chips and scratches, finish problems caused by rail brake dust and upholstery stains are the most common problems. Major claims for collision damage, broken windshields and suspension problems occur infrequently. Figure A.1 indicates that most claims can be traced to carriers.

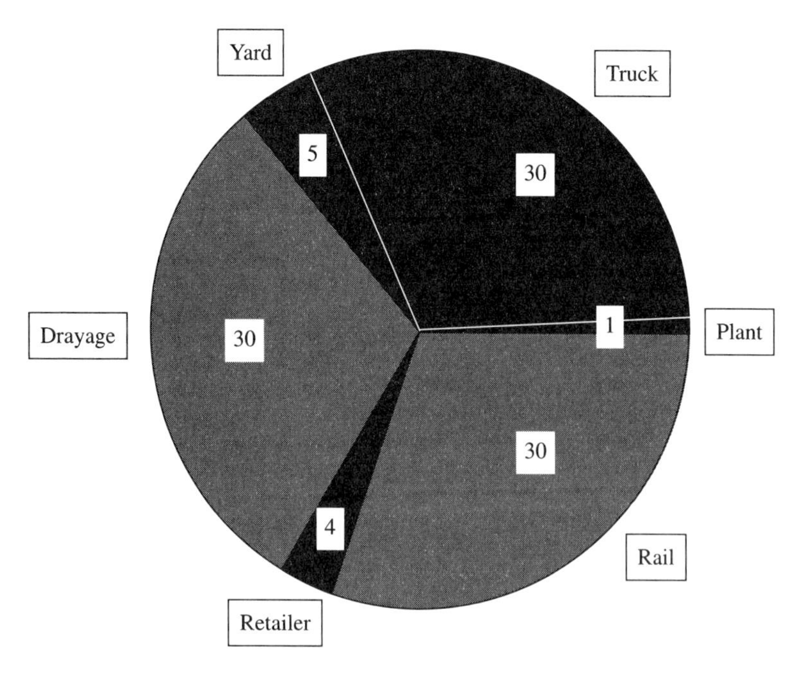

Figure A.1 Responsibility for vehicle damage.

Transit time

Transit time measurements focus on the interval from event 42 (vehicle dispatched) to event 50 (vehicle received by retailer). Truck transit time begins when the load is released by QAT and clears the Saturn security gate. Rail transit time begins when CSX pulls a block of railcars off the Saturn property. Transit time ends when vehicles are available for retailer inspection. On-time percentage, transit time average and transit time variation are analysed for both modes.

Truck delivery is much faster and more consistent than is rail delivery. This should come as no surprise given the disparity of delivery ranges (10–500 miles for truck compared with 500–2500 miles for rail). The truck transit time average is 1.3 days, with a standard deviation of 0.9 days. On-time performance is 98%–99% for truck deliveries. In contrast, the rail transit time average is 7.6 days, with a standard deviation of 2.3 days. On-time performance ranges from 75%–80% for rail deliveries.

Analytical approach

The pros and cons for the two delivery alternatives suggest that there is no simple solution to this problem. Each location must be reviewed individually based on the cost and service levels highlighted in Table A.2, the potential for damage claims and the other major stakeholder issues discussed throughout the case. Thus the challenge at hand is to analyse all relevant selection factors in an objective, systematic manner before making each delivery decision.

Given the type of information available an appropriate analytical approach would be a categorical rating system. The current and anticipated performance of the two delivery methods in each key area would be evaluated via a simple grading system, such as 'satisfactory', 'neutral' and 'unsatisfactory'. These grades could then be assigned scores of +1, 0 and −1 respectively, for developing a composite evaluation of carrier performance. This approach is inexpensive, requires minimal performance data and lends structure to an otherwise intuitive process. With this approach in mind and an array of performance information in hand, the Vehicle Transportation Team can proceed with its task of selecting the best delivery method to serve each new Saturn retailer.

A.1.5 Case questions

A.1.1 Identify the primary issues related to the selection of outbound transportation providers and discuss their importance to each process stakeholder. What other factors and issues should the Vehicle Transportation Team consider in the selection process?

A.1.2 Starting with event 38, develop flowcharts of the delivery process for truck and rail. Develop a second set of flowcharts that streamlines the delivery processes, removes redundant, unimportant activities and addresses concerns related to the current process.

A.1.3 Which delivery method would you recommend for each new retailer to minimize the outbound logistics costs per vehicle?

A.1.4 Which delivery method would you recommend for each new retailer to minimize transit time and transit time consistency?

A.1.5 Assuming that each new retailer will sell 720 vehicles annually, which delivery method would you recommend for each new retailer to minimize vehicle damage?

A.1.6 What is your final recommendation (truck or rail delivery) for each new retailer? Why? Which stakeholder groups are likely to be most or least satisfied with your decision? In addition, map the US territory that is likely to be served by truck delivery.

A.1.7 CSX has suggested that Saturn consolidate all deliveries for southeast Georgia and the east coast of Florida for a once per week unit train delivery. The total outbound cost would be $210 per vehicle and CSX guarantees a 3 day transit time. How would this affect Saturn's delivery methods to their retailers in Savannah, GA, and Jacksonville, FL?

A.1.8 Burlington Northern Santa Fe (BNSF) Railroad has approached Saturn about using the Autostack system to deliver vehicles. Autostack is an intermodal delivery system in which six Saturns can be loaded into a standard 48 ft container at Spring Hill and delivered to destination railyards via doublestack intermodal service. The cars would be unloaded and delivered by drayage carriers. BNSF has offered 2 day transit time to Arlington, TX, at a rate of $232 per vehicle. Loading costs of $13.50 and drayage costs of $12.50 would be incurred. How would this opportunity affect the delivery method recommendation for the Dallas, TX, retailer?

A.2 CASE 2. TOYS 'R' US JAPAN

This case describes the growth of Toys 'R' Us (TRU) as the leading retailer of toys in the United States, its international expansion and its specific entry into the problematic Japanese market. Certain 'non-market' or politically-based obstacles had to be overcome to enter the Japanese market, primarily the Daitenho or 'Big Store Law.' Once it entered, TRU Japan proved that its retail format was successful in the Japanese market. However, this initial success does not guarantee its future. Large discount stores, increasing since the reform of the Daitenho, are rapidly providing competition that could eventually threaten TRU Japan's expansion and diminish future profitability.

Although there appears to be a strong sales potential, TRU Japan is still far from developing a toy store network comparable to that in the United States. TRU Japan needs to expand rapidly, developing a distribution and logistics network that will enable TRU to capture the efficiencies which are vital in sustaining its low-price, selection and in-stock strategy that it has employed successfully in the United States. TRU appears to be making headway, but, as the case ends, management needs to respond quickly to the crisis of losing its warehouse and distribution center in Kobe.

A.3 CASE 2A. TOYS 'R' US JAPAN

Tuesday, 6 January 1992, Toys 'R' Us (TRU) held its grand opening in Kashihara, Nara-ken, Japan. Arriving by helicopter, US President George Bush appeared at the opening ceremonies for the second TRU store in Japan. Attending were Minister Kozo Watanabe of Japan's Ministry of International Trade and Industry (MITI), the US Commerce Secretary Robert Mosbacher, US Ambassador to Japan Michael Armacost, Japanese Ambassador to the United States Ryohei Murata, Toys 'R' Us Chairman Charles Lazarus and the local governor and mayor. About 2000 Nara-ken police and students from local police academies were mobilized in a massive security measure. About 5000 people came to witness the event; many of them waved small Japanese and US flags .

President Bush thanked the gathered officials and praised the progress of the Structural Impediments Initiative (SII) to remove economic barriers to trade, create more jobs in the United States and bring the Japanese consumer world-class goods. He continued:

> 'And what makes me so happy here today is that we see here the beginning of a dynamic, new economic relationship. One of greater balance. There is much that we can do for the world based on a forward-looking global partnership between two great nations, two powerful economies and two resourceful innovative peoples'.

Few Americans would attach a great deal of significance to the public appearance of politicians at ceremonies such as these. Surprising to TRU, however, the showcase appearance of the US President at Nara had an immediate impact upon Japanese toy industry vendors in the weeks following the event. Negotiations suddenly improved between TRU Japan and its Japanese suppliers. Japanese vendors apparently began to take seriously the previous announcement that TRU Japan intended to open 100 stores in the next 10 years and become the major toy retailer in Japan.

A.3.1 Company background

The first TRU store was opened in Washington, DC, in 1957 by Charles Lazarus. Three stores opened over the next 10 years and Lazarus sold his ownership stake for $7.5 million to Interstate Stores in 1966. When problems with other Interstate divisions drove the corporation into bankruptcy proceedings, Lazarus regained control in 1978 through a management-led buyout. The TRU strategy is based upon price, selection and keeping stores in stock. As Lazarus explained, 'When a customer walks through our doors with a shopping list, we better have 95% of what's on her list or we're in trouble.' The EDLP (every day low prices) strategy and in-stock image stimulates purchasing year-round instead of primarily during the Christmas season. Baby diapers and formula are sold at or below cost in the hope of winning over new parents and keeping them as customers as their children mature. This strategy has won TRU a steadily increasing share of the retail toy market, rising to about 22% in 1995.

TRU shifted its goals for expansion dramatically in 1983. The firm entered the children's clothing market with Kids 'R' Us and established the International Division. Joseph Baczko was recruited from his job as chief executive of the European operations of Max Factor to lead the international expansion. Baczko perceived that there were increasing global opportunities in the toy business. In an article in 1986 for the industry trade magazine, *Playthings*, he noted that customers overseas had higher disposable income, were more educated and had more free time. Moreover, these buyers were more price-conscious and tended to prefer specialty retailers – factors that favored the international expansion of TRU.

The first international store opened in 1984 in Canada. In 1986 TRU struck joint venture deals in Singapore and Hong Kong. The company next expanded to the United Kingdom in 1987, into Germany in 1988, and into France and Taiwan in 1989. By 1994 TRU had penetrated the Nordic countries and had developed new franchise relationships with Top-Toy A/S, the leading Scandinavian toy retailer. The franchise division also led to the entry of TRU to Israel, Saudi Arabia and the United Arab Emirates, markets which would otherwise be prohibitive because of cultural differences and restrictive laws.

TRU learned to adapt to the different competitive retail situations in each country that it entered. Different countries can have drastically different competitive environments. For example, supermarket toy sales as a percentage of all toy sales range from about 4% in the United Kingdom to 48% in France. High costs in land, labor and distribution created problems in maintaining the TRU price and selection strategy. Low-cost retail sites proved difficult to find in England, leading TRU to try smaller store formats. In Germany, competing retailers initially pressured vendors not to sell to TRU. Nevertheless, sales increased and store expansion was rapid. Even in England, where British parents spend less on toys, the number of shoppers per store was very high. New store openings attracted 40 000 shoppers in Hong Kong and 20 000 to a one-acre site in Frankfurt. International sales grew to about one-quarter of company revenues by 1994.

Customer preferences can vary enormously among countries, hence TRU had to control its product mix carefully. Porcelain dolls are carried in Japan, whereas Germans prefer wooden dolls. TRU sells a version of Monopoly in Hong Kong that replaces 'Boardwalk' and 'Park Place' with 'Sheko' and 'Repulse Bay,' and those in France stock scale models of the French high-speed train. Whereas about 70%–80% of its European toy sales are the same items as those in the United States, in Japan this number is only about 30%–40%.

TRU has constantly worked to refine the warehouse toy store concept in its home market. To service customers better, TRU linked store managers' pay to customer service activities and tried 'store within a store' or 'boutique' concepts. These included Lego shops, 'plush' or Stuffed Animal shops, Learning Centers and entertainment software sections. The most successful of these was the book department, called Books 'R' Us, a joint effort with Western Publishing that requires special chairs, lights, carpets and tables. Though it may not have as wide a selection of books as bookstores, the book department enables TRU to pick up sales from parents supplementing a toy purchase with a book having better 'educational' value.

As a specialty store 'category killer,' TRU competes across retail categories, not just with toy stores. To compete better with discounters, in 1994 TRU developed coupon books to offer deeper price savings. A 'Big Toy Book Catalogue' a 'Video Game & Electronic Toys Catalogue' and an internet connection were introduced as shopping aids. Additionally, the 'Toy Guide for Differently Abled Kids' offers professional evaluations of toys for families of disabled children; this complements the TRU corporate giving program which focuses on improving the health-care needs of children.

A.3.2 Going to Japan

Japan was a particularly attractive target for TRU for several reasons. Japan was the second largest toy market in the world. The 'Statistics of Toys' of the Japan Toy Association estimated the size of the overall Japanese toy market on a retail price basis at ¥932 to ¥950 billion in 1991. By the 1980s Japan had developed a particularly high per capita income, and toys sales had been growing despite the low birthrate. Spending on children was particularly high, particularly in the early years of childhood, yet toy shops in Japan tended to have small selections and high prices.

As TRU formulated plans to enter the Japanese market it first had to deal with the barrier posed by local laws and politics. The biggest impediment was the Daitenho or Big Store law, which went into effect in 1974. The law stated that local store owners must give their consent before a retail outlet with floor space larger than roughly 5400 ft^2 can be opened in their locale. Under the guise of protecting the community, the law granted small local businesses the authority to force new competitors into a review process which could last more than 10 years. The Daitenho effectively enabled local shopkeepers to block the establishment of TRU stores in Japan beginning in late 1988.

Though originally intended to protect the small shops, the Daitenho effectively made the retail network of stores rather rigid, benefiting the large incumbent stores established before the law was created. Store chains such as Yaohan could not expand in Japan, and Yaohan eventually moved its headquarters to Hong Kong to concentrate upon international expansion. Opponents of the Daitenho included Shinji Shimiju, chief executive officer of the Lion supermarket convenience store chain, who sued MITI in the mid-1980s on the basis that the law barred competition. Lawsuits are uncommon in Japan, but press coverage on the suit helped focus public attention to the fact that chains such as Daiei and Ito-Yokado were unfairly benefiting from the law.

Widespread publicity about the TRU difficulties to open new stores gained attention in both the US and the Japanese press. Being unable to get approval for any of its stores, TRU appealed for help directly through the US Trade Representative and other channels. Lobbying efforts by Japanese business had been taking place for several years to change the Daitenho, and US negotiation pressure reinforced this effort. In 1989 TRU International Division president Joseph Baczko met Takuro Isoda of Daiwa Securities America while sitting on Georgetown University's forum on Japan. Isoda introduced Baczko to Den Fujita, President of McDonald's Japan, which by that time had already developed a network of 700 shops yielding $1.2 billion in sales. Their compatibility led McDonald's Japan to take a 20% stake in the TRU Japan subsidiary. Den Fujita had substantial experience with real estate and numerous government contacts, many of them fellow alumni of the University of Tokyo's law department.

MITI had shown it was receptive to the relaxation of the Daitenho in the spring of 1989 in a report it circulated entitled *Vision of the Japanese Retail Industry in the 1990s*. As Kabun Muto rose to become the Minister of MITI he encouraged the idea of formally revising the Daitenho, not just altering MITI's interpretation of the law. As discussions were taking place within the Japanese government, Fujita traveled to the United States to meet with Trade Representative Carla Hills in 1990. Fujita urged her office to take diplomatic efforts to push for changes in the Daitenho. Fujita knew that the time was right for a change in the law. MITI was ready to be persuaded by the Americans to take action publicly.

A.3.3 Japan: distribution and retail environment

Japanese society is hierarchical and group-orientated; there is a strong emphasis on maintaining harmony and avoiding confrontations. Sensitivity to social practices is of great importance in establishing successful business relationships in Japan. The view of the individual differs significantly in comparison with European or North American standards. It is important to acknowledge and respect obligations, created through educational ties, employment, favors or assistance. Obligations tend to be mutually felt and bind business partners to each other. Failure to repay a perceived debt or recompense for an obligation can bring about a 'loss of face' for a person who perceives himself or herself as indebted. Since business in Japan tends to be conducted on the basis of long-term relationships and trust, entry into the Japanese market generally requires a long-term commitment. Proper introductions and personal contacts can be highly important. This has fueled the claim that the Japanese business system gives inordinate advantages to entrenched incumbents to the detriment of newcomers. The web of overlapping relationships which constitutes the distribution sector of the economy is particularly difficult for foreign firms to penetrate.

In addition, many Japanese business sectors, such as telecommunications, financial services and construction, remain heavily regulated. In the retail sector, laws restrict the size of buildings that retailers can erect and dictate store layouts and construction materials. Rules and regulations apply to all companies, Japanese included, but government ties and relationships can sometimes give established firms an edge in getting past the red tape. With 124 million people residing in a mountainous country, 70% of Japan's population is concentrated in coastal areas constituting about 20% of the country. Densely populated Japanese cities have high real-estate prices. As a consequence, homes and stores are small. Japanese households have small kitchen areas with little room to store fresh foods, and shopping tends to occur on a daily basis.

The Japanese distribution system involves many middle men and includes 1.6 million 'mom and pop' stores, half of which sell only food items. The average Japanese retail store has only about 3200 ft^2 of floor space and limited storage capacity. Distributors make small deliveries of less than full case quantities. Japanese consumers tend to make frequent small purchases from shopowners who they know within their neighborhood setting, particularly in the food sector. Small shops account for 56% of retail sales in Japan as compared with 3% for the

United States and 5% for Europe. The number of retail outlets in Japan is nearly the same as that in the United States, despite the fact that its population is roughly half that of the United States and that Japan is slightly smaller than California.

Mom and pop stores are closely connected with neighborhood and community events. Every shopping street or district in Japan has its own retailer's union which puts up seasonal decorations in the streets, holds store-front sales and organizes special events. Most events coincide with *matsuri* (festivals) such as summer festivals in August, the hottest month of the year, that can draw substantial crowds. Some groups run drawings or coupon shopping schemes that are aimed at encouraging local residents to patronize their neighborhood stores. Cultural values can serve to reinforce the complex network of wholesalers. Small stores may not be able to secure credit by themselves and often are provided financial, ownership or other exclusive arrangements with major Japanese manufacturers, industrial groups or trading companies. Japanese wholesalers have become powerful through their capacity to provide financial support to smaller manufacturers and retailers who have limited resources. Having limited space and high storage costs, retailers rely on wholesalers to maintain inventories, obligating them further to wholesalers. Since one of every five Japanese workers is employed in distribution, change is likely to entail political problems.

A.3.4 Changing the Daitenho

The rising level of wealth in the 1980s permitted many more Japanese to travel outside the country than before. Exposed to new cultures, Japanese consumers began to question high prices, government regulation and distribution practices within Japan. Popular opinion toward many traditional aspects of business changed. As Japan entered the 1990s it was troubled by economic and political problems. The so-called 'bubble economy' of the 1980s burst in 1991, plunging Japan into a recession. The Liberal Democratic Party (LDP), which had been in control since the early 1950s, began to lose its grip on power as a result of scandals amongst its leadership. A parade of 'revolving door' prime ministers even included the advancement of the Socialist party leader, Tomichi Murayama, to the Prime Minister's office in 1994.

The Structural Impediments Initiative (SII), a set of discussions between the United States and Japan, was launched in July 1989 to address the underlying causes of the bilateral trade imbalance, examining macroeconomic policies and business practices as trade barriers. In June 1990 the two sides released a report in which they made commitments to reduce structural impediments and to meet for follow-up discussions from 1990 to 1993. Japan's multilayered distribution sector became an issue in these discussions. MITI agreed, in April 1990, to limit to 18 months the application process under the Daitenho. This 18 month period included six months for consultations with local small storeowners, eight months for negotiating details with those stores and four months for adjustments. Further changes came on 31st January 1992 as MITI reduced the waiting period to 12 months and upgraded membership of the local 'advisory councils.' These councils reviewed applications for new stores and could demand certain reductions in store size.

For type 1 large stores, which include supermarkets and department stores, the maximum size was enlarged to 3000 m^2 (about 30 000 ft^2) from the previous 1500 m^2. For type 2 specialty stores such as TRU, the maximum size was enlarged from 500 m^2 to 3000 m^2. Exceptions to size regulations applied to those stores opening in Tokyo's 23 wards and in major cities, where the maximum was set to 6000 m^2. Finally, on 1st May 1994 MITI made two further concessions. Large stores were allowed to extend their closing time to 8 p.m., and the number of days stores were required to close for holidays was reduced to 24, from 44.

A.3.5 Getting started in Japan

TRU Japan established an office in Kawasaki and targeted development of store sites in Niigata, Chiba, Sagamihara and Fukuoka. Although MITI had agreed to revisions to shorten the approval process, TRU also had to deal with local and regional ordinances. For example, certain changes in the Daitenho agreed to by MITI required the approval of the Ministry of Home Affairs. In January 1990 TRU Japan submitted the paperwork to open a store in Niigata on the northern coast of Japan. To win local approval from the Niigata Chamber of Commerce and Industry, TRU agreed to shrink the size of its originally planned store by 30% and the opening became further delayed.

In May 1990 TRU Japan started talks on opening a store in Sagamihara, a Tokyo suburb with a population of 520 000, submitting applications to the ministry, Kanagawa Prefecture, Sagamihara City and the Chamber of Commerce. In August, TRU participated in an 'explanation meeting' at the local public hall and later made presentations to the Sagamihara Commercial Activities Council, a MITI-inspired body of 18 consumers, merchants, professionals and academics. After four meetings, the council gave approval in June 1991 to allow TRU to open its store some time after 1st December. The requirement was that, like other stores, TRU had to close every day by 8 p.m. and not open on at least 30 days of the year. In addition, TRU had to consult with other regulatory bodies over possible traffic congestion problems. By September, those consultations and construction delays forced a postponement of the Sagamihara opening until March 1992. Determined to open one store in 1991, TRU tried to hasten a similar approval process of its planned store in Ibaraki.

Meanwhile, other obstacles had surfaced. Takashigi Seki, of the group called the All Japan Associated Toys, cited with alarm to the press that British toy stores had decreased from 6000 to 1500 after the TRU entry. In Japan, the total number of toy stores, which had peaked at about 8000, had recently declined to about 6000. Next, in September 1990, 520 small toy retailers formally announced forming the Japan Association of Specialty Toy Shops, to 'protect its members from the likes of TRU Japan' by lobbying and other means. Public statements indicated that the association intended to provide members with research on toy retailing, employee education, sales promotion and management techniques. A total of 11 major retailers led the lobbying association, including Kiddy Land in Tokyo, Pelican in Osaka and Angel in Fukuoka. Toshikazu Koya, leader of the association, practicing attorney and president of the 53-store Kiddy Land chain, explained that he planned to pool purchases with other Japanese retailers to buy large lots of inventory at competitive prices.

Japanese toy retailers and wholesalers bitterly complained that TRU would put them out of business. In seeking retail space to lease, TRU found that some were reluctant to lease TRU space which could upset local business clients. Most seriously, toy vendors, worried about their long-established relationships with toy retailers, refused to sell to TRU or deal directly with them. TRU vice chairman, Robert Nakasone, himself an American of Japanese descent, confidently remarked to the press that disputes with Japanese toy makers could be resolved. TRU had been dealing with many Japanese companies for years. Large toy manufacturers such as Nintendo, having established a long-term business relationship with TRU, could not 'lose face' by refusing to do business with TRU in Japan. Eventually, Nintendo publicly announced their intention to supply directly TRU Japan, negotiating prices that would not offend other retailers. Other manufacturers began to follow.

On its part, however, TRU was committed to working with local suppliers. One reason for this was to stock toys carrying the ST mark. The ST or Safety Toy mark is a voluntary quality mark that can be applied to toys that meet the safety standards set by the Japanese toy industry through the Toy Safety Control Administration, a self-regulatory commission composed of toy manufacturers, consumers and health professionals. The Japan Toy Association established the ST mark and had long engaged in information campaigns aimed at educating Japanese consumers about the merits of buying toys affixed with the ST mark. Over 95% of toys sold in Japan, whether domestic or imported, carry the ST mark.

While cooperating with Japanese vendors in many respects, TRU rebelled against the Japanese practice of selling at manufacturers' suggested retail prices. TRU maintained that prices should not be decided by either manufacturers or wholesalers and initiated direct dealings with manufacturers. For the most part, TRU was forced to go through established distribution channels. Although TRU was a large distributor in the United States, this had little importance to most Japanese toy vendors since it was regarded as a small company within Japan. TRU moved forward with its plans aggressively. As Joseph Baczko left to become chief executive officer of Blockbuster, Larry Bouts was brought in from Pepsico Foods to head the TRU International Division. Den Fujita announced to the press a plan to link McDonald's Japan with TRU Japan and Blockbuster in a large suburban store concept, calling the group the MTB Rengo. With a bit of marketing flourish, Fujita referred to the plan as the 'Meiji Ishin of distribution', comparing the idea to the Japanese political revolution at the end of the samurai period.

On 20th December 1991, TRU Japan opened its first retail store in Japan in Amimachi, Ibaraki-ken, about 40 miles northeast of Tokyo. With the colorful, English-language TRU sign in front of the store and an 850-car parking lot, the store is similar to those in the United States. It is smaller than was hoped, with retail floor space of about 3000 m^2 or 32 400 ft^2 as opposed to the average of 46 000 ft^2 in the United States – but still, this is about 10 times larger than the average toy store in Japan. Whereas the typical small Japanese toy store stocks between 1000 and 2000 different items, TRU Japan started out with about 8000 and eventually increased this number to 15 000 items. With all the publicity that TRU had gotten in the newspapers, a crowd of 17 000 jammed the store on opening day and set a TRU grand-opening sales record.

A.3.6 Case questions

A.3.1 What advantages did TRU have that helped it to enter the Japanese market?

A.3.2 How did the 20% stake taken by McDonald's help TRU to gain entry?

A.3.3 What problems do you anticipate that TRU will have in the next few years as it expands?

A.4 CASE 2B. TOYS 'R' US JAPAN

The date was 16th January 1995, the place Toys 'R' Us (TRU) Headquarters, Paramus, NJ. Managers at TRU had been assessing sales

Table A.4 Consolidated statement of earnings. Source: Toys 'R' Us Annual Report

	Earnings ($ millions)		
	28 Jan. 1995	*29 Jan. 1994*	*30 Jan. 1993*
Net sales	8746	7946	7169
Cost of sales	6008	5495	4968
Selling, advertising, general and administration expense	1664	1497	1342
Operating income	1074	954	859
Other income	16	24	19
Total income	1090	978	878
Depreciation and amortization	161	133	119
Interest expense	84	73	69
Interest and other income	16	24	19
Earnings before taxes on income	844	773	689
Income tax	312	290	252
Net earnings	532	483	438
Earnings per share ($)	1.85	1.63	1.47

and inventory levels in newly-opened retail stores around the world (Table A.4). With 48% of annual TRU sales occurring in the last quarter, the year had come to an end with the usual frenzied spree of last-minute Christmas buying. Sales figures were being analysed to aid buying decisions for the next month's American International Toy Fair, the large trade show of over 1600 exhibitors held in New York every year. Sales of video games in the United States continued to be weak, as customers awaited the new generation of 32 and 64 bit systems. TRU Japan, part of the International Division, could provide sales information on the 32 bit systems of Sega and Sony already introduced in Japan, where new product introductions had boosted performance in the fourth quarter. Analysis of sales in Japan could aid the development of sales plans that would more accurately set inventory levels in the US and other markets. For the past 12 years the International Division had contributed to the organization in several ways – it had just achieved a 37% increase in operating earnings as it improved upon inventory management and increased productivity in labor and distribution.

As TRU expanded rapidly in diverse markets in Europe, Asia and Australia (Tables A.5 and A.6), it faced new problems and competitors. TRU had to address unique country problems with suppliers, local regulations and interest groups to meet demand for toys globally. Company policies in warehousing and inventory management also needed to be assessed at each location. Japan was proving to be a particularly successful market – 13 new stores were scheduled to open in 1995 to bring the total count to 37. Entry had been difficult and with its high costs, traditional business culture and new generation of

Table A.5 Selected store data by year. Source: Toys 'R' Us (TRU) Annual Report

Fiscal year ended	*TRU (USA)*	*Kids 'R' Us*	*International*	*Number of Countries*	*Japan*	*Warehouses*	*Net Sales ($ millions)*	*Net Earnings ($ millions)*	*Number Times Earnings*
28 Jan. 1995	618	204	293	21	37	15	8746	532	10.21
29 Jan. 1994	581	217	234	20	24	17	7946	483	10.70
30 Jan. 1993	540	211	167	16	16	na	7169	438	9.86
1 Feb. 1992	497	189	126	11	11	na	6124	340	8.53
2 Feb. 1991	451	164	97	10	6	na	5510	326	7.21
28 Jan. 1990	404	137	74	8	1	21	4788	321	10.57
29 Jan. 1989	358	112	52	8		21	4000	268	11.39
31 Jan. 1988	313	74	37	6		20	3137	204	16.39
1 Feb. 1987	271	43	24	5		18	2445	152	20.80
2 Feb. 1986	233	23	13	4		14	1976	120	20.38
1985	198		5	3		14			
1984	169		0	1		13			
1983	144		0			12			
1982	120		0			11			
1981	101		0			10			

Table A.6 Number of stores, by location. Source: Toys 'R' Us (TRU) Annual Report

Country	*Number of Stores*	*Country*	*Number of Stores*
Australia	17	Malaysia	3
Austria	7	The Netherlands	8
Belgium	3	Portugal	3
Canada	56	Singapore	3
Denmark	1	Spain	20
France	29	Sweden	3
Germany	53	Switzerland	4
Hong Kong	4	Taiwan	4
Japan	24	United Arab Emirates	1
Luxembourg	1	United Kingdom	49

Total:

International TRU	293	(337[a])
US TRU[b]	618	(653[a])
US Kids 'R' Us[c]	206	(213[a])

[a] End of 1995.
[b] Includes 48 states and 4 in Puerto Rico.
[c] Includes 28 states, serviced from three distribution centers.

consumers Japan continued to pose some particularly difficult management problems. Later in the evening, TRU Japan had some disastrous news for the home office – there were problems at the distribution center in Kobe.

A.4.1 The toy business

Although there are many traditional best-selling toys and games, many toys have a limited popularity (Table A.7) and appeal in different ways

Table A.7 Top 10 toys of 1994. Source: *Playthings*, 1994, Annual Survey of US Buyers

Rating	*Game (Manufacturer)*
1	Mighty Morphin Power Rangers (Bandai America)
2	Barbie (Mattel)[a]
3	The Lion King (Mattel)
4	Genesis (Sega)[a]
5	Batman (Kenner)[a]
6	Jennie Gymnast (Mattel)
7	Super Nintendo Entertainment System (Nintendo)[a]
8	GI Joe (Hasbro)[a]
9	Bumble Ball (Ertl)
10	Cool Tools (Playskool)

[a] Appeared on the list in 1993.

Table A.8 Percentage of toy sales per age group, 1991. Source: The NPD Group, Port Washington, NY

Age group	*Percentage sales*
up to 11 months	4.0
12–23 months	7.1
2–3 years	14.4
4–5 years	16.1
6–7 years	14.5
8–9 years	11.1
10–12 years	13.9
13–17 years	6.9
18 plus[a]	12.4

[a] Includes collectibles. All figures include video games.

among the various children's age segments (Table A.8). While retailers may sell toy products in every toy category, market shares vary considerably on an individual product category basis (Table A.9). When a new toy or game based upon characters in television or movies becomes successful, companies license its name, character or logo, bringing about even greater exposure. With about 40% of retail toy sales from new products, everyone in the toy business is constantly looking for hits. Yet failures are common – only about 100 of the 2000 new toys introduced each year make it to a second year.

Since parents pay close attention to how toys affect their children, the publicity a toy receives can be very important. Toy stores removed realistic-looking toy guns from their shelves when, in two separate incidents, police mistook toy guns as real and shot two children. Besides guns and war toys, video games also have been criticized as being violent. Responding to public complaints, TRU stopped selling the Sega Genesis game *Night Trap* in 1993.

Recent consolidations among toy manufacturers have created sizable companies. In less than a decade, Hasbro acquired Milton Bradley, Playskool, Tonka, Kenner, Parker Brothers, Coleco, Knickerbocker and Child Guidance, growing to about 15% of the US toy market sales. Mattel's $1.2 billion purchase of Fisher Price in 1993 made it second largest. These companies, together with Tyco Toys, The Lego Group and Little Tikes (a unit of Rubbermaid), accounted for about 60% of the industry's business in 1994.

Toy manufacturers broaden their product line to increase their influence with retailers. The five big retailers in the toy business – TRU, Wal-Mart, Kmart, Target Stores (a unit of Dayton Hudson) and Kay-Bee Toy Stores (a unit of Melville) – sell approximately 60% of all toys in the United States. Manufacturers use the few hits they have as leverage to get retailers to accept their full line of merchandise. Retailers, on

Table A.9 Distribution of toy category sales percentages by outlet. Source: The NPD Group, Port Washington, NY, Toy Market Index, 1993–1995 *Playthings*

	Plush		*Construction Toys*	*Family Action Games*	*Children's Action Games*	*Family Board Games*	*Children's Board Games*	*Non-Powered Trucks*	*Radio Controlled Toys (1994)*	*Wood or Plastic Puzzles*	*Cardboard Puzzles*
	1992	*1994*									
Discount stores	33	35	56	49	45	44	47	51	40	42	44
Toys 'R' Us stores	7	8	21	27	17	22	23	12	26	6	15
Other national toy chains	4	0	4	8	7	6	6	2	0	6	8
All other toy stores	5	11	4	2	3	5	5	2	5	5	6
Department stores	12	11	4	3	3	4	4	3	6	4	4
Variety stores	3	4	1	0	8	3	3	4	1	6	6
Catalog showrooms	1	2	2	4	3	2	2	1	4	0	2
Food and drug	8	11	1	0	5	2	3	4	2	15	2
All other outlets	27	22	7	7	9	12	7	21	16	16	13
Unit sales (thousands)	71 306	128 324	39 100	2004	24 553	7331	15 239	19 494	11 097	57 861	5112
Sales ($ thousands)	716 321	1197 865	510 579	30 537	209 644	99 227	167 340	222 404	368 836	196 917	28 387

Note: all figures are for 1993, except where stated otherwise.

their part, seek such things from manufacturers as volume rebates, advertising allowances, and credits for store displays, and this may lead to certain conflicts. For example, Kmart buyers asked about 20 of its 200 toy suppliers to sell their goods on consignment in 1993 (meaning that Kmart would not pay for the toys until they appeared on store shelves). Most toy manufacturers refuse to sell on consignment.

Since a large portion of the toy business is done during the short, high-demand Christmas buying season, it is important that retailers anticipate demand and stay in stock. Since the majority of toy consumers are children, sales trends are difficult to research. Sales of particular toys may be virtually impossible to predict. Difficulties in meeting demand for highly popular toys has occurred many times in the past. Severe shortages occurred for Cabbage Patch dolls in 1983, Ninja Turtles in 1987 and Power Rangers in 1993. Forecasting errors can be expensive. For example, Worlds of Wonder raised $80 million in a bond offering in July of 1987 but by Christmas the company was in bankruptcy – the majority of the money raised was tied up in inventory. Bankruptcies of several toy manufacturers occurred throughout the 1980s, including Coleco, the original maker of the Cabbage Patch dolls.

With uncertain demand, retailers try to limit their risk by closely monitoring and controlling inventories. As stores move closer to just-in-time distribution, limited orders are placed and shipments are timed to replace products as they are taken off the shelf. The problem is that sales rates vary enormously. Sales indicators of what may be the hot-selling Christmas items may not be seen until late September or October – by that time it may be too late to reorder and expect shipments to arrive before Christmas. Retailers may be pitted against each other to have their orders filled quickly by vendors. Since it takes an average of three months to make, ship and stock a product, manufacturers and retailers can face extremely difficult decisions about inventory levels.

A.4.2 Logistics and expansion

Efficient operations form a vital component in maintaining the TRU price and selection strategy. TRU was among the first in the retail industry extensively to employ computer technology in managing inventories and in using automation to support distribution. Utilizing better inventory tracking systems and more efficient distribution networks allowed TRU to reduce the 21 distribution centers down to 15, even as the sales volume grew significantly. Two automated distribution centers replaced four older facilities in 1994.

Inventory is tracked in its movement throughout the supply chain, with effective communication being a key to control inventory levels and monitor costs. Careful, fast replenishment of store inventory diminishes the likelihood of keeping warehouses full of unwanted products. 'The later you flow the goods through the channels, the lower you keep your inventory and the fewer your markdowns,' says Michael Goldstein, vice chairman of TRU. 'Electronic tracking keeps TRU battle-station ready.' From computer command posts, 'we know what's on our trucks, where they're headed, and who needs new supplies.'

As TRU has grown it has developed increasingly efficient and sophisticated systems. Sales are tracked with computers linked to cash-register scanners via satellite. This permits sales trends to be monitored immediately and store inventory quickly replenished. TRU is able to communicate via Hughes VSAT (very small aperture terminals) satellite service, most of its stores having a satellite dish located on the roof. The VSAT connects to each store's Ethernet LAN. TRU has tightly integrated communication between vendors, stores and the mainframe at headquarters. The network gives store managers fast access to inventory data and provides quicker verification of credit accounts. The bandwidth that VSAT utilizes also allows for broadcast television, making possible televised employee training programs, companywide announcements and other corporate communications.

TRU hired the logistics unit of American President Company (APC), the large West Coast steamship and train operator, to control and track the movement of shipments from Asian factories. APC offers an advanced system of high-tech hardware and customized software to monitor closely the progress of 8000 40-foot containers shipping toys from the Far East. About 2000 of these move through Port Newark and Miami. Containers are routed to go direct to regional distribution centers. This saves time and space, avoiding shipments from being unpacked in West Coast warehouses and re-sorted for shipment to retail outlets. The APC system provides TRU with precise, timely information on the location and contents of every container at any given time. Maureen Saul, director of traffic at TRU, notes that getting such information early 'can literally be as important as the actual movement of the goods.' Receipt of information up to two weeks before the cargo arrives in the United States enables the staff to reorder missing products, inform stores and customers of potential delays and, if necessary, divert cargo en route.

Efficient operations saves the company millions of dollars by reducing carrying costs, minimizing markdowns and avoiding additional handling expenses. This is vital for cost-competitive pricing and limiting the risks of carrying a wide selection of merchandise. As TRU improved efficiency and perfected its control systems in the United States, it brought considerable advantages to competing in international markets.

A.4.3 Managing store expansion in Japan

TRU Japan stores operate similarly to those in the United States. Clerks are few and hard to find and checkout is highly mechanized. One difference, however, is that shoplifting is so rare that there are no hidden security cameras. Few other theft-prevention measures are needed. In general, when purchases are made in Japan, stores will typically carefully dust each item, remove its price tag and wrap it in attractive paper. At TRU Japan stores, however, clerks simply place the item into a plastic bag and move on to the next customer. This policy appeared to matter very little to Japanese customers who appreciated the low prices and large selections. Sales continued to grow at each new TRU Japan store.

Prices at TRU Japan were initially only about 10% lower than those in competing stores, forcing TRU to compete more on the basis of its wide selection and in-stock supply. TRU Japan was committed to lowering prices, however, and worked on developing greater efficiency and obtaining better price deals from vendors. As TRU opened new stores in Sapporo and other locations, management tried to locate in low-rent locations and employed extensive part-time help to keep costs minimal.

Whereas in the United States automated systems at distribution centers could sort and label merchandise at high speeds better to meet peaks in demand, such efficiencies would only be possible in Japan if TRU Japan grew to a sufficient size to justify the cost. Innovations such as the satellite communications systems, which could be readily connected in each new US store, were prohibitively expensive to the small network of stores in Japan. TRU Japan had to find new efficiencies and develop its distribution network. TRU Japan directed its effort toward continuous expansion, from Sapporo on the northern island of Hokkaido, to Okinawa in the far south. Several problems had to be met to make this expansion successful.

1. Stores had to be placed at locations to build a sufficient degree of customer traffic while keeping costs low. Options were limited – while suburban stores could be located in proximity to roads and highways which could make them visible, accessible by car and available to truck supply routes, TRU also needed to consider that many Japanese consumers travel primarily by train.
2. TRU had to deal with the costs of construction, estimated to be about three times higher than the United States. With careful selection of material, sources and suppliers, TRU Japan developed a way to complete construction for about half the cost normally required in Japan.
3. TRU had to plan carefully for increases in supply and distribution capacity as new stores came on line. This was complicated by the

uncertainties of the store approval process. TRU filed applications and timed the forwarding of government documents and paperwork to match the growing capacity of its developing store network.

A.4.4 Operations in Japan

Stores generated considerable customer traffic and produced sales from $15 million to $20 million (compared with $10 million for the typical US store). Each new and popular TRU store convinced building owners that other businesses could benefit from the TRU presence, giving TRU more opportunities and leverage. The continuing recession also led to a fall in land prices, yielding greater flexibility in site selection. As recessionary conditions worsened, Japanese manufacturers had even more impetus to offer better deals. Though it was a continuous struggle to buy direct from Japanese manufacturers, TRU eventually managed to cut direct purchase deals with more than 50 Japanese toy makers.

Training was also a challenge. Ken Bonning, vice-president of International Distribution at TRU, noted that although Japanese managers can be good at accurately inspecting and managing the flow of paper documents, the 'paperless' computerized inventory management system at TRU Japan represented something of a challenge to traditional methods. Finding employees experienced in computerized systems was a problem, often making training in the United States a necessity. TRU recruited young Japanese people who were educated in the United States to train on running its computerized systems. In addition, managers from the United States were sent to Japan to train and develop staff that could respond to changing toy demand in the Japanese market.

Stores in Japan are run on the 'push' system which TRU developed in the United States. In this system store managers are responsible for the cost of operating their individual stores. The head office in Kawasaki is responsible for buying and inventory planning decisions. Japanese buyers make product selection decisions and formulate detailed sales plans for their market. These plans are used as the basis for planning cost-efficient transportation schedules which are revised as necessary to meet demand. Given the fluctuations in demand in the toy business, managers need constantly to monitor positive and negative sales trends and respond to inventory levels appropriately.

A.4.5 The supply chain in Japan

When TRU entered Japan it had the opportunity to create goodwill among US toy manufacturers by distributing their products in Japan. Not only did TRU select products for the Japanese market and provide vendors with shelf space, but TRU also provided sales information necessary to support and develop the sales of their products. TRU could not only offer greater product variety in its stores, but it also created opportunities for manufacturers abroad to grow and explore new options.

TRU set up its own direct import company to assist efficient deliveries straight from the docks. Japanese customs officials at first insisted upon tough inspection standards; for example when trace amounts of formaldehyde (one part per billion) were found in a pricing sticker, merchandise was rejected. Such unrealistic standards were mitigated over time with assistance from the US consul; it later became a routine process to get a toy approved. Japanese officials eventually agreed to other reasonable changes, for example to allow TRU to send toys to foreign laboratories to be tested for Japanese safety standards and to scrap the requirement that foreign toys be retested every six months.

Further store openings enabled TRU Japan to increase volume, and this allowed TRU to negotiate better price and delivery terms. As the distribution channels which the vendors were using began to have difficulty handling large volumes, Japanese toy vendors began to suggest that TRU deal with them directly. Although TRU had always sought direct purchase deals, these were somewhat easier to accomplish with the larger toy manufacturers such as Bandai. Bandai had bought out some of its distributors in an effort to reduce its costs.

TRU did not get all that it wanted from its vendors, however. TRU Japan had little control over its inbound logistics; vendors insisted that they alone handle deliveries to the TRU Japan warehouse. Since a significant portion of toy merchandise from Japanese vendors is imported from southeast Asian manufacturers, this prevented TRU from consolidating shipments or pursuing other transportation efficiencies. The warehouse decision was a difficult one – Japanese consider this work *'kitsui, kitanai,* and *kiken'* – labor intensive, dirty and dangerous. After considering several locations for its growing network, TRU Japan decided to locate its distribution center in the Kansai area at Kobe. Kobe offered a large, modernized port, a centralized location, was less expensive in terms of labor costs and was close to large urban centers including Osaka and Kyoto. Leasing a warehouse location at Nada put it in proximity to the Hanshin Expressway. This was advantageous in that the expressway was equipped with a unique automated traffic control system. The system was linked with detectors every 800 m which minimized traffic congestion problems.

The growing network of stores continued to expand (Table A.10). TRU Japan was rapidly achieving the critical mass that allowed for increased distribution and logistical efficiencies. A second warehouse and distribution center was planned for Yokohama, the large port city near Tokyo. As more stores came on line, however, new competitive

Table A.10 Toys 'R' Us Japan: the first 24 store locations

1. Arakawaoki, Ami-Machi, Inashiki-Gun, Ibaraki
2. Chiba-Chuo, Chiba-Shi Chiba
3. Hakata, Fukuoka-Shi Fukuoka
4. Himeji, Himeji-Shi Hyogo
5. Ichikawa, Ichikawa-Shi, Chiba
6. Kadena, Kadena-Cho, Nakakami-Gun, Okinawa
7. Kashihara, Kashihara-Shi, Nara
8. Kuki, Kuki-Shi, Saitama
9. Miyakojima, Miyakojima-Ku, Osaka-Shi, Osaka
10. Nagano, Nagano-Shi, Nagano
11. Nagasaki, Nagasaki-Shi Nagasaki
12. Natori, Miyagi
13. Niigata, Niigata-Shi, Niigata
14. Nishikasugai, Nishikasugai-Gun Aichi
15. Ohtsu, Ohtsu-Shi, Shiga
16. Okayama, Okayama-Shi, Okayama
17. Okazaki, Okazaki-Shi, Aichi
18. Sagamihara, Sagamihara-Shi, Kanagawa
19. Sapporo-Hassamu, Sapporo-Shi, Hokkaido
20. Sendai-Izumi, Sendai-Shi, Miyagi
21. Shingu, Kasuya-Gun, Fukuoka
22. Suminoe-Koen, Osaka-Shi Osaka
23. Wakayama, Wakayama-Shi Wakayama
24. Yaenosato, Higashiosaka-Shi, Osaka

problems emerged. The growing number of discounters who were selling toys, increasing since the change in the Daitenho, was creating competition for TRU Japan. One positive effect was that, with the growth of many large retailers, opposition to TRU seemed to disappear. Faced with price declines, retailers and wholesalers in many retail categories were forced to streamline their distribution systems further. New store formats and the growth of discounters continued to bring important changes to retail and distribution sectors in Japan.

A.4.6 The crisis

At 5:46 a.m. on Tuesday 17th January 1995 a massive earthquake hit the Kobe area. Measuring 7.2 on the Richter scale, the earthquake rocked the densely populated region, toppled buildings, sheared major highways and rail lines and spread fires throughout the city. The losses associated with the quake were staggering: 5500 lives, 350 000 injuries, 300 000 left homeless, over 180 000 buildings destroyed and an estimated $147 billion in direct damages.

Japanese government authorities were unprepared to cope with the disaster, local authorities admitting that disaster plans had never been considered. In the weeks following the quake stories spread of how government officials took days to take action. No clear lines of authority for disaster relief had been established that would permit effective coordination between central, regional and local government authorities. A strong sense of national pride and self-sufficiency prompted Japanese officials to reject at least half the offers of aid from over 70 international organizations around the world. Risk-averse Japanese bureaucrats resisted allowing medicine and relief personnel into the country without subjecting them to time-consuming procedures. Some of these requirements were later eased, but too late to make a significant difference to the victims.

In the absence of response by the government, groups and individuals took the initiative and became sources of vital food and water supplies. Companies such as Daiei and Seven-Eleven of Japan responded quickly and effectively to the victims. Yakuza (gangsters) transported relief supplies such as water, food, toiletries and diapers into the area and freely distributed them to local residents. One estimate has it that large enterprises in the quake zone lost 30% of their productive capacity and smaller enterprises 50%. Substantial indirect costs were also incurred as a result of impediments to distribution and lack of back-up or contingency plans. Few companies had a crisis management manual or one that was of any real use. The earthquake severely damaged the Port of Kobe, the sixth largest container port in the world, and greatly hampered ground transportation. Of eight major transportation routes that ran through Kobe, only one remained operational. This greatly impeded relief efforts and crippled trade.

The TRU warehouse was in the center of one of the worst hit areas. Even if workers were safe and the products could be recovered, roads were clogged with rubble and a major section of the Hanshin Expressway nearby had crumbled. The warehouse would not be operative. As they learned of the quake, managers at the TRU Kawasaki office quickly began to consider the alternatives. At this critical stage in the growth of TRU Japan, the loss of the warehouse in Kobe would pose some difficult challenges.

A.4.7 Case questions

A.4.1 Explain the role of logistics in carrying out the international expansion strategy of TRU. Be sure to consider factors such as store size, forecasting, information systems, competition as well as distribution networks in your answer.

A.4.2 How does the relationship between TRU and its suppliers differ in the United States and Japan? How does the expansion of TRU into international markets affect their relationships?

A.4.3 How does the TRU network of stores differ between the United States and Japan? How would this affect the ability of TRU to supply its stores efficiently?

A.4.4 What suggestions do you have for TRU Japan to respond to the Kobe warehouse disaster?

FURTHER READING

Winter, D. (1995) Saturn Turns 10. *Ward's Auto World*, 67–71.

Bradley, P. (1994) Logistics joins links in automotive supply chain. *Purchasing*, 57–9.

John, F. and Taguchi, F. (1995) Reassessing the Japanese distribution system. *Sloan Management Review*, 49–61.

McGinnis, M.A. (1990) The relative importance of cost and service in freight transportation choice: before and after deregulation. *Transportation Journal*, 12–19.

Yoshi, T. (1990) *Global Management: Business Strategy and Government Policy*, Copley, Acton, MA.

Index